MW00980872

Geographic Information Science and Systems

Geographic Information Science and Systems

Alexa Franklin

www.clanryeinternational.com

Clanrye International,
750 Third Avenue, 9th Floor,
New York, NY 10017, USA

ISBN: 978-1-64726-135-1

Cataloging-in-Publication Data

Geographic information science and systems / Alexa Franklin.
p. cm.
Includes bibliographical references and index.
ISBN: 978-1-64726-135-1
1. Geographic information systems. 2. Information storage and retrieval systems--Geography. I. Franklin, Alexa.
G70.212 .G46 2021
910.285--dc23

For information on all Clanrye International publications
visit our website at www.clanryeinternational.com

Table of Contents

Preface

This book has been written, keeping in view that students want more practical information. Thus, my aim has been to make it as comprehensive as possible for the readers. I would like to extend my thanks to my family and co-workers for their knowledge, support and encouragement all along.

Geographic information system is the system that is specifically designed to capture, store, analyze, manage and present spatial and geographical data. GIS applications function as tools which allow users to create interactive queries, edit data in maps, analyze spatial information and present the results of such operations. The science underlying the geographic concepts, applications and systems is termed as geographic information science. Modern GIS technologies make use of digital information, which is obtained through various digitized data creation methods. The most common method of data creation is digitization in which a hard copy map is transferred into a digital medium by a CAD program and geo-referencing capabilities. This book provides comprehensive insights into the field of geographic information system. Most of the topics introduced herein cover new techniques and the applications of this field. For someone with an interest and eye for detail, this book covers the most significant topics in this area.

A brief description of the chapters is provided below for further understanding:

Chapter - What is Geographic Information System?

A system which is designed to capture, manipulate, store, analyse, present and manage spatial or geographic data is referred to as geographic information system. It plays an integral role in the study of the spatial patterns and relationships. This chapter has been carefully written to provide an easy introduction to the varied facets of geographic information systems

Chapter - Components of Geographic Information System

Some of the components of geographic information system are object based spatial data models, geospatial data models, geodata and geometric networks. The diverse applications of these components as well the major concepts related to geographic information system such as GIS mapping and geospatial intelligence have been thoroughly discussed in this chapter.

Chapter - Techniques, Tools and Software in Geographic Information System

Various kinds of techniques, tools and software are used in geographic information system. Some of these are remote sensing, hydrological modeling, cartographical modeling, geoprocessing tools, spatial join tool, etc. This chapter discusses in detail these techniques, tools and software related to geographic information system.

Chapter - Spatial Analysis in Geographic Information System

The set of formal techniques that study entities using their geometric, geographic and topological properties are referred to as spatial analysis. The chapter closely examines the key concepts related to spatial analysis such as spatial data and its types provide an extensive understanding of the subject.

Chapter - Understanding Geocoding

The process of transforming a description to a location on the surface of the Earth is known as geocoding. Some of the major concepts studied in relation to geocoding are geocodes, geocoding addresses, geolocation and reverse geocoding. This chapter has been carefully written to provide an easy understanding of these aspects of geocoding.

Chapter - Applications of Geographic Information System

The geographic information systems are applied in various areas. Some of these include industries, archaeology, public health, aquatic science, environment, Google maps, Google Earth, geospatial intelligence, etc. This chapter discusses in detail these diverse applications of geographic information system.

Alexa Franklin

1

What is Geographic Information System?

A system which is designed to capture, manipulate, store, analyse, present and manage spatial or geographic data is referred to as geographic information system. It plays an integral role in the study of the spatial patterns and relationships. This chapter has been carefully written to provide an easy introduction to the varied facets of geographic information systems.

A geographic information system (GIS) is a computer-based tool for mapping and analyzing things that exist and events that happen on Earth. GIS technology integrates common database operations such as query and statistical analysis with the unique visualization and geographic analysis benefits offered by maps. These abilities distinguish GIS from other information systems and make it valuable to a wide range of public and private enterprises for explaining events, predicting outcomes, and planning strategies.

Map making and geographic analysis are not new, but a GIS performs these tasks better and faster than do the old manual methods. And, before GIS technology, only a few people had the skills necessary to use geographic information to help with decision making and problem solving.

Components of a GIS

A working GIS integrates five key components: hardware, software, data, people, and methods.

Hardware

Hardware is the computer on which a GIS operates. Today, GIS software runs on a wide range of hardware types, from centralized computer servers to desktop computers used in stand-alone or networked configurations.

Software

GIS software provides the functions and tools needed to store, analyze, and display geographic information. Key software components are:

- Tools for the input and manipulation of geographic information
- A database management system (DBMS)
- Tools that support geographic query, analysis, and visualization
- A graphical user interface (GUI) for easy access to tool.

Data

Possibly the most important component of a GIS is the data. Geographic data and related tabular data can be collected in-house or purchased from a commercial data provider. A GIS will integrate spatial data with other data resources and can even use a DBMS, used by most organizations to organize and maintain their data, to manage spatial data.

People

GIS technology is of limited value without the people who manage the system and develop plans for applying it to real world problems. GIS users range from technical specialists who design and maintain the system to those who use it to help them perform their everyday work.

Methods

A successful GIS operates according to a well-designed plan and business rules, which are the models and operating practices unique to each organization.

How GIS Works?

A GIS stores information about the world as a collection of thematic layers that can be linked together by geography. This simple but extremely powerful and versatile concept has proven invaluable for solving many real-world problems from tracking delivery vehicles, to recording details of planning applications, to modeling global atmospheric circulation.

Geographic References

Geographic information contains either an explicit geographic reference such as a latitude and longitude or national grid coordinate, or an implicit reference such as an address, postal code, census tract name, forest stand identifier, or road name. An automated process called geocoding is used to create explicit geographic references (multiple locations) from implicit references. These geographic references allow you to locate features such as a business or forest stand and events such as an earthquake on the Earth's surface for analysis.

Vector and Raster Models

Geographic information systems work with two fundamentally different types of geographic models--the "vector model" and the "raster model."

In the vector model, information about points, lines, and polygons is encoded and stored as a collection of x,y coordinates. The location of a point feature, such as a bore hole, can be described by a single x,y coordinate. Linear features, such as roads and rivers, can be stored as a collection of point coordinates. Polygonal features, such as sales territories and river catchments, can be stored as a closed loop of coordinates. The vector model is extremely useful for describing discrete features, but less useful for describing continuously varying features such as soil type or accessibility costs for hospitals.

The raster model has evolved to model such continuous features. A raster image comprises a collection of grid cells rather like a scanned map or picture. Both the vector and raster models for storing geographic data have unique advantages and disadvantages. Modern GISs are able to handle both models.

GIS Tasks

GISs essentially perform five processes or tasks:

- Input
- Manipulation
- Management
- Query and Analysis
- Visualization

Input

Before geographic data can be used in a GIS, the data must be converted into a suitable digital format. The process of converting data from paper maps into computer files is called digitizing. Modern GIS technology has the capability to automate this process fully for large projects using scanning technology; smaller jobs may require some manual digitizing.

Today many types of geographic data already exist in GIS-compatible formats. These data can be obtained from data suppliers and loaded directly into a GIS.

Manipulation

It is likely that data types required for a particular GIS project will need to be transformed or manipulated in some way to make them compatible with your system. For example, geographic information is available at different scales (street centerline files might be available at a scale of 1:100,000; census boundaries at 1:50,000; and postal codes at 1:10,000). Before this information can be integrated, it must be transformed to the same scale. This could be a temporary transformation for display purposes or a permanent one required for analysis. GIS technology offers many tools for manipulating spatial data and for weeding out unnecessary data.

Management

For small GIS projects it may be sufficient to store geographic information as simple files. There comes a point, however, when data volumes become large and the number of data users becomes more than a few, that it is best to use a database management system (DBMS) to help store, organize, and manage data. A DBMS is nothing more than computer software for managing a database -an integrated collection of data.

There are many different designs of DBMSs, but in GIS the relational design has been the most useful. In the relational design, data are stored conceptually as a collection of tables. Common fields in different tables are used to link them together. This surprisingly simple design has been so widely used primarily because of its flexibility and very wide deployment in applications both within and without GIS.

- Query and Analysis

Once you have a functioning GIS containing your geographic information, you can begin to ask simple questions such as:

- Who owns the land parcel on the corner?
- How far is it between two places?
- Where is land zoned for industrial use?

And analytical questions such as:

- Where are all the sites suitable for building?
- What is the dominant soil type for oak forest?
- If a new highway is built here, how will traffic be affected?

GIS provides both simple point-and-click query capabilities and sophisticated analysis tools to provide timely information to managers and analysts alike. GIS technology really comes into its own when used to analyze geographic data to look for patterns and trends, and to undertake "what if" scenarios. Modern GISs have many powerful analytical tools, but two are especially important.

- Proximity Analysis:
    - How many houses lie within 100 m of this water main?
    - What is the total number of customers within 10 km of this store?
    - What proportion of the alfalfa crop is within 500 m of the well?

To answer such questions, GIS technology uses a process called buffering to determine the proximity relationship between features.

- Overlay Analysis

The integration of different data layers involves a process called overlay. At its simplest, this could be a visual operation, but analytical operations require one or more data layers to be joined physically. This overlay, or spatial join, can integrate data on soils, slope, and vegetation, or land ownership with tax assessment.

Visualization

For many types of geographic operation the end result is best visualized as a map or graph. Maps are very efficient at storing and communicating geographic information. While cartographers have created maps for millennia, GIS provides new and exciting tools to extend the art and science of cartography. Map displays can be integrated with reports, three-dimensional views, photographic images, and other output, such as multimedia.

What can GIS do?

Perform Geographic Queries and Analysis

The ability of GIS to search databases and perform geographic queries has saved many companies literally millions of dollars. GISs have helped:

- Decrease the time taken to answer customer requests.
- Find land suitable for development.
- Search for relationships among crops, soils, and climate.
- Locate the position of breaks in electrical circuits.

A realtor could use a GIS to find all houses within a certain area that have tiled roofs and five bedrooms, then list their characteristics. You could also list houses within a certain distance of a school.

Improve Organizational Integration

Many organizations that have implemented a GIS have found that one of its main benefits is improved management of their own organization and resources. Because GISs have the ability to link data sets together by geography, they facilitate interdepartmental information sharing and communication. By creating a shared database one department can benefit from the work of another--data can be collected once and used many times.

As communication increases among individuals and departments, redundancy is reduced, productivity is enhanced, and overall organizational efficiency is improved. Thus, in a utility company the customer and infrastructure databases can be integrated so that when there is planned maintenance, affected customers can be sent a computer-generated letter.

Make Better Decisions

The old adage "better information leads to better decisions" is as true for GIS as it is for other information systems. A GIS, however, is not an automated decision making system but a tool to query, analyze, and map data in support of the decision making process. GIS technology has been used to assist in tasks such as presenting information at planning inquiries, helping resolve territorial disputes, and siting pylons in such a way as to minimize visual intrusion.

GIS can be used to help reach a decision about the location of a new housing addition that has minimal environmental impact, is located in a low risk area, and is close to a population center. The information can be presented succinctly and clearly in the form of a map and accompanying report, allowing decision makers to focus on the real issues rather than trying to understand the data. Because GIS products can be produced quickly, multiple scenarios can be evaluated efficiently and effectively.

Make Maps

Maps have a special place in GIS. The process of making maps with GIS is much more flexible than are traditional manual or automated cartography approaches. It begins with database creation.

Existing paper maps can be digitized and computer-compatible information can be translated into the GIS. The GIS-based cartographic database can be both continuous and scale free. Map products can then be created centered on any location, at any scale, and showing selected information symbolized effectively to highlight specific characteristics.

The characteristics of atlases and map series can be encoded in computer programs and compared with the database at final production time. Digital products for use in other GISs can also be derived by simply copying data from the database. In a large organization, topographic databases can be used as reference frameworks by other departments.

Pros and Cons of Geographic Information System

Advantages of GIS

- Geographic information systems can visualize spatial information.
- It can be used for a vast range of tasks involving geography.
- It can provide the solutions for the problems and it can model natural disaster activity precisely.
- Gis technology offers time management.
- It offers a quick collection of data.
- It presents a catalog of data.
- Display spatial information (Spatial = Graphics + Tables)
- It has high accuracy and presents better predictions and analysis.
- It has the ability of improving the organizational integration which helps software to talk each other.
- Gis would also allow viewing, questioning, understanding, visualizing and interpreting the data into number of ways which will reveal relationships, trends and patterns in the form of globes, maps, charts and reports.
- A gis helps you answer the questions and solve the problems by analyzing your data and outputed in a easy and meaningful way.
- Gis data is used in the natural resource management that can include hillslope gradients, aspect, stream network, stream gradient, slope, catchment area and more.

Disadvantages of GIS

- Geographic Information System (GIS) is very expensive software.
- It requires the enormous amount of data inputted to be practical for task so there is changes of error.
- It has relative resolution loss.

- It has a privacy violation.
- Geographic Information System's error increases for larger scale map because of the earth shape (not perfect round).
- Funding for GIS is needed because it is more costly.
- If GIS is run by non-GIS group, Geography might suffer.
- GIS layers may cause some costly mistakes when it is handled by non-GIS person.
- The data is fuel for GIS, if there is no data then GIS cannot function.
- GIS output will be accurate if the input Data is correct, so GIS Expert has to be alwasys aware of fact that all data are not always correct.
- There is always a failure to fully complete the GIS implementation in an organization.
- It's very hard to make GIS programs which are both fast and user-friendly, GIS systems typically require complex command language.
- Data fields, and their accessibility are not very understood and data can become incomplete, obsolete or erroneous, rendering the GIS misleading.

DISTRIBUTED GIS

Distributed GIS refers to GI Systems that do not have all of the system components in the same physical location. This could be the processing, the database, the rendering or the user interface. Examples of distributed systems are web-based GIS and Mobile GIS.

Types

Enterprise GIS

Enterprise GIS refers to a geographical information system that integrates geographic data across multiple departments and serves the whole organisation. The basic idea of an enterprise GIS is to deal with departmental needs collectively instead of individually. When organisations started using GIS in the 1960s and 1970s, the focus was on individual projects where individual users created and maintained data sets on their own desktop computers. Due to extensive interaction and work-flow between departments, many organisations have in recent years switched from independent, stand-alone GIS systems to more integrated approaches that share resources and applications.

Some of the potential benefits that an enterprise GIS can provide include significantly reduced redundancy of data across the system, improved accuracy and integrity of geographic information, and more efficient use and sharing of data. Since data is one of the most significant investments in any GIS program, any approach that reduces acquisition costs while maintaining data quality is important. The implementation of an enterprise GIS may also reduce the overall GIS maintenance and support costs providing a more effective use of departmental GIS resources. Data can be integrated and used in decision making processes across the whole organisation.

Corporate GIS

A corporate Geographical Information System, is similar to Enterprise GIS and satisfies the spatial information needs of an organisation as a whole in an integrated manner. Corporate GIS consists of four technological elements which are data, standards, information technology and personnel with expertise. It is a coordinated approach that moves away from fragmented desktop GIS. The design of a corporate GIS includes the construction of a centralised corporate database that is designed to be the principle resource for an entire organisation. The corporate database is specifically designed to efficiently and effectively suit the requirements of the organisation. Essential to a corporate GIS is the effective management of the corporate database and the establishment of standards such as OGC for mapping and database technologies.

Benefits include that all the users in the organisation have access to shared, complete, accurate, high quality and up-to-date data. All the users in the organisation also have access to shared technology and people with expertise. This improves the efficiency and effectiveness of the organisation as a whole. A successfully managed corporate database reduces redundant collection and storage of information across the organisation. By centralising resources and efforts, it reduces the overall cost.

Mobile GIS

With ~80% of all data deemed to have a spatial component, modern Mobile GIS are a powerful geo-centric business process integration platform enabling the Spatial Enterprise. The number of mobile devices in circulation has surpassed the world's population (2013) with a rapid acceleration in iOS, Android and Windows 8 tablet up-take. Tablets are fast becoming popular for Utility field use. Low-cost MIL-STD-810 certified cases transform consumer tablets into fully ruggedised, yet lightweight field use units at 10% of legacy ruggedised laptop costs.

Although not all applications of mobile GIS are limited by the device, many are. These limitations are more applicable to smaller devices such as cell phones and PDAs. Such devices have: small screens with a poor resolution, limited memory and processing power, a poor (or no) keyboard, and short battery life. Additional limitations can be found in web client based tablet applications: poor web GUI and device integration, on-line reliance, and very limited off-line web client cache.

Location-based Services

Location-based services (LBS) are services that are distributed wirelessly and provide information relevant to the user's current location. These services include such things as 'find my nearest', directions, and various vehicle monitoring systems, such as the GM OnStar system amongst others. Location-based services are generally run on mobile phones and PDAs, and are intended for use by the general public more than Mobile GIS systems which are geared towards commercial enterprise. Devices can be located by triangulation using the mobile phone network and GPS.

Web Mapping Services

A web mapping service is a means of displaying and interacting with maps on the Web. The first web mapping service was the Xerox PARC Map Viewer built in 1993 and decommissioned in 2000.

There have been 3 generations of web map service. The first generation was from 1993 onwards and consisted of simple image maps which had a single click function. The second generation was from 1996 onwards and still used image maps the one click function. However, they also had zoom and pan capabilities (although slow) and could be customised through the use of the URL API. The third generation was from 1998 onwards and were the first to include slippy maps. They utilise AJAX technology which enables seamless panning and zooming. They are customisable using the URL API and can have extended functionality programmed in using the DOM.

Web map services are based on the concept of the image map whereby this defines the area overlaying an image (e.g. GIF). An image map can be processed client or server side. As functionality is built into the web server, performance is good. Image maps can be dynamic. When image maps are used for geographic purposes, the co-ordinate system must be transformed to the geographical origin to conform to the geographical standard of having the origin at the bottom left corner.

Web maps are used for location-based services.

Local Search

Local Search is a recent approach to internet searching that incorporates geographical information into search queries so that the links that you return are more relevant to where you are. It developed out of an increasing awareness that many search engine users are using it to look for a business or service in the local area. Local search has stimulated the development of web mapping, which is used either as a tool to use in geographically restricting your search or as an additional resource to be returned along with search result listings. It has also led to an increase in the number of small businesses advertising on the web.

Mashups

In distributed GIS, the term mashup refers to a generic web service which combines content and functionality from disparate sources; mashups reflect a separation of information and presentation. Mashups are increasingly being used in commercial and government applications as well as in the public domain. When used in GIS, it reflects the concept of connecting an application with a mapping service. An examples is combining Google maps with Chicago crime statistics to create the Chicago crime statistics map. Mashups are fast, provide value for money and remove responsibility for the data from the creator.

Second generation systems provide mashups mainly based on URL parameters, while Third generation systems (e.g. Google Maps) allow customisation via script (e.g. JavaScript).

Strategy

The development of the European Union (EU) Infrastructure for Spatial Information in the *European Community* (INSPIRE) initiative indicates this is a matter that is gaining more awareness at the national and EU scale. This states that there is a need to create 'quality geo-referenced information' that would be useful for a better understanding of human activities on environmental processes. Therefore, it is an ambitious project that aims to develop a European spatial information database.

The GI strategy for Scotland was introduced in 2005 to provide a sustainable SDI, through the "One Scotland – One Geography" implementation plan. This documentation notes that it should be able to provide linkages to the "Spaces, Faces and Places of Scotland". Although plans for a GI strategy have been in existence for some time, it was revealed at the AGI Scotland 2007 conference that a recent budget review by the Scottish Government indicated there will not be an allocation of resources to fund this initiative within the next term. Therefore, a business plan will need to be presented in order to outline the cost-benefits involved with taking up the strategy.

Standards

The main standards for Distributed GIS are provided by the Open Geospatial Consortium (OGC). OGC is a non-profit international group which seeks to Web-Enable GIS and in turn Geo-Enable the web. One of the major issues concerning distributed GIS is the interoperability of the data since it can come in different formats using different projection systems. OGC standards seek to provide interoperability between data and to integrate existing data.

OGC

In terms of interoperability, the use of communication standards in Distributed GIS is particularly important. General standards for Geospatial Data have been developed by the Open Geospatial Consortium (OGC). For the exchange of Geospatial Data over the web, the most important OGC standards are Web Map Service (WMS) and Web Feature Service (WFS).

Using OGC compliant gateways allows for building very flexible Distributed GI Systems. Unlike monolithic GI Systems, OGC compliant systems are naturally web-based and do not have strict definitions of servers and clients. For instance, if a user (client) accesses a server, that server itself can act as a client of a number of further servers in order to retrieve data requested by the user. This concept allows for data retrieval from any number of different sources, providing consistent data standards are used. This concept allows data transfer with systems not capable of GIS functionality. A key function of OGC standards is the integration of different systems already existing and thus geo-enabling the web. Web services providing different functionality can be used simultaneously to combine data from different sources (mash-ups). Thus, different services on distributed servers can be combined for 'service-chaining' in order to add additional value to existing services. Providing a wide use of OGC standards by different web services, sharing distributed data of multiple organisations becomes possible.

Some important languages used in OGC compliant systems are described in the following. XML stands for eXtensible Markup language and is widely used for displaying and interpreting data from computers. Thus the development of a web-based GI system requires several useful XML encodings that can effectively describe two-dimensional graphics such as maps SVG and at the same time store and transfer simple features GML. Because GML and SVG are both XML encodings, it is very straightforward to convert between the two using an XML Style Language Transformation XSLT. This gives an application a means of rendering GML, and in fact is the primary way that it has been accomplished among existing applications today. XML can introduce innovative web services, in terms of GIS. It allows geographic information to be easily translated in graphic and in these terms scalar vector graphics (SVG) can produce high quality dynamic outputs by using data retrieved from spatial databases. In the same aspect Google, one of the pioneers in web-based GIS, has developed its own language which also uses a XML structure. Keyhole Markup Language (KML) is a file format used to display geographic data in an earth browser, such as Google Earth, Google Maps, and Google Maps for mobile browsers *"Google KML definition"*.

Global System for Mobile Communications

Global System for Mobile Communications (GSM) is a global standard for mobile phones around the world. Networks using the GSM system offer transmission of voice, data and messages in text and multimedia form and provide web, telenet, ftp, email services etc. over the mobile network. Almost two million people are now using GSM. Five main standards of GSM exist: GSM 400, GSM 850, GSM 900, GSM-1800 (DCS) and GSM1900 (PCS). GSM 850 and GSM 1900 is used in North America, parts of Latin America and parts of Africa. In Europe, Asia and Australia GSM 900/1800 standard is used.

GSM consists of two components: the mobile radio telephone and Subscriber Identity Module. GSM is a cellular network, which is a radio network made up of a number of cells. For each cell, the transmitter (known as a base station) is transmitting and receiving signals. The base station is controlled through the Base Station Controller via the Mobile Switching Centre.

For GSM enhancement General Packet Radio Service (GPRS), a packet-oriented data service for data transmission, and Universal Mobile Telecommunications System (UTMS), the Third Generation (3G) mobile communication system, technology was introduced. Both provide similar services to 2G, but with greater bandwidth and speed.

Wireless Application Protocol

Wireless Application Protocol (WAP) is a standard for the data transmission of internet content and services. It is a secure specification that allows users to access the information instantly via mobile phones, pagers, two-way radios, smartphones and communicators. WAP supports HTML and XML, and WML language, and is specifically designed for small screens and one-hand navigation without a keyboard. WML is scalable from two-line text displays up to the graphical screens found on smart phones. It is much stricter than HTML and is similar to JavaScript.

Geotagging

Geotagging is the process of adding geographical identification metadata to resources such as websites, RSS feed, images or videos. The metadata usually consist of latitude and longitude coordinates but may also include altitude, camera holding direction, place information and so on. Flickr website is one of the famous web services which host photos and provides functionality to add latitude and longitude information to the picture. The main idea is to use metadata related to pictures and photo collection. A geotag is simply a properly-formed XML tag giving the geographic coordinates of a place. The coordinates can be specified in latitude and longitude or in UTM (Universal Transverse Mercator) coordinates.

The RDFIG Geo vocabulary from the W3C is the common basis for the recommendations. It supplies official global names for the latitude, longitude, and altitude properties. These are given in a system of coordinates known as "the WGS84 datum". A geographic datum specifies an ellipsoidal approximation to the Earth's surface; WGS84 is the most commonly used such datum.

Parallel Processing

Parallel processing is the use of multiple CPU's to execute different sections of a program together. Remote sensing and surveying equipment have been providing vast amounts of spatial information, and how to manage, process or dispose of this data have become major issues in the field of Geographic Information Science (GIS). To solve these problems there has been much research into the area of parallel processing of GIS information. This involves the utilization of a single computer with multiple processors or multiple computers that are connected over a network working on the same task. There are many different types of distributed computing, two of the most common are clustering and grid processing.

Some consider grid computing to be "the third information technology wave" after the Internet and Web, and will be the backbone of the next generation of services and applications that are going to further the research and development of GIS and related areas. Grid computing allows for the sharing of processing power, enabling the attainment of high performances in computing, management and services. Grid computing, (unlike the conventional supercomputer that does parallel computing by linking multiple processors over a system bus) uses a network of computers to execute a program.

The problem of using multiple computers lies in the difficulty of dividing up the tasks among the computers, without having to reference portions of the code being executed on other CPUs. Amdahl's law expresses the speedup of a program as a result of parallelization. It states that potential program speedup is defined by the fraction of code (P) that can be parallelized: $1/(1-P)$. If the code cannot be broken up to run over multiple processors, $P = 0$ and the speedup = 1 (no speedup). If it is possible to break up the code to be perfectly parallel then $P = 1$ and the speedup is infinite, in theory though practical limits occur. Thus, there is an upper bound on the usefulness of adding more parallel execution units. Gustafson's law is a law closely related to Amdahl's law but doesn't make as many assumptions and tries to model these factors in the representation of performance. The equation can be modelled by $S(P) = P - \alpha * (P - 1)$ where P is the number of processors, S is the speedup, and α the non-parallelizable part of the process.

PUBLIC PARTICIPATION GEOGRAPHIC INFORMATION SYSTEM

A public participation geographic information system (PPGIS) is meant to bring the academic practices of GIS and mapping to the local level in order to promote knowledge production by local and non-governmental groups. The idea behind PPGIS is empowerment and inclusion of marginalized populations, who have little voice in the public arena, through geographic technology education and participation. PPGIS uses and produces digital maps, satellite imagery, sketch maps, and many other spatial and visual tools, to change geographic involvement and awareness on a local level. The term was coined in 1996 at the meetings of the National Center for Geographic Information and Analysis (NCGIA).

Applications

Attendees to the Mapping for Change International Conference on Participatory Spatial Information Management and Communication conferred to at least three potential implications of PPGIS; it can: (1) enhance capacity in generating, managing, and communicating spatial information; (2) stimulate innovation; and ultimately; (3) encourage positive social change.

There are a range of applications for PPGIS. The potential outcomes can be applied from community and neighborhood planning and development to environmental and natural resource management. Marginalized groups, be they grassroots organizations to indigenous populations could benefit from GIS technology.

Governments, non-government organizations and non-profit groups are a big force behind many programs. The current extent of PPGIS programs in the US has been evaluated by Sawicki and Peterman. They catalog over 60 PPGIS programs who aid in "public participation in community decision making by providing local-area data to community groups," in the United States. The organizations providing these programs are mostly universities, local chambers of commerce, non-profit foundations.

In general, neighborhood empowerment groups can form and gain access to information that is normally very easy for the official government and planning offices to obtain. It is easier for this to happen than for individuals of lower-income neighborhoods just working by themselves. There have been several projects where university students help implement GIS in neighborhoods and communities. It is believed that access to information is the doorway to more effective government for everybody and community empowerment. In a case study of a group in Milwaukee, residents of an inner city neighborhood became active participants in building a community information system, learning to access public information and create and analyze new databases derived from their own surveys, all with the purpose of making these residents useful actors in city management and in the formation of public policy. In many cases, there are providers of data for community groups, but the groups may not know that such entities exist. Getting the word out would be beneficial.

Some of the spatial data that the neighborhood wanted was information on abandoned or boarded-up buildings and homes, vacant lots, and properties that contained garbage, rubbish and debris that contributed to health and safety issues in the area. They also appreciated being able to find landlords that were not keeping up the properties. The university team and the community were able to build databases and make maps that would help them find these areas and perform the spatial analysis that they needed. Community members learned how to use the computer resources, ArcView 1.0, and build a theme or land use map of the surrounding area. They were able to perform spatial queries and analyze neighborhood problems. Some of these problems included finding absentee landlords and finding code violations for the buildings on the maps.

Approaches

There are two approaches to PPGIS use and application. These two perspectives, top–down and bottom–up, are the currently debated schism in PPGIS.

Top-down

According to Sieber, PPGIS was first envisioned as a means of mapping individuals by many social and economic demographic factors in order to analyze the spatial differences in access to social services. She refers to this kind of PPGIS as *top-down*, being that it is less hands on for the public, but theoretically serves the public by making adjustments for the deficiencies, and improvements in public management.

Bottom-up

A current trend with academic involvement in PPGIS, is researching existing programs, and or starting programs in order to collect data on the effectiveness of PPGIS. Elwood in *The Professional Geographer*, talks in depth about the "everyday inclusions, exclusions, and contradictions of Participatory GIS research." The research is being conducted in order to evaluate if PPGIS is involving the public equally. In reference to Sieber's top-down PPGIS, this is a counter method of PPGIS, rightly referred to as *bottom-up* PPGIS. Its purpose is to work with the public to let them learn the technologies, then producing their own GIS.

Public participation GIS is defined by Sieber as the use of geographic information systems to broaden public involvement in policymaking as well as to the value of GIS to promote the goals of nongovernmental organizations, grassroots groups and community-based organizations. It would seem on the surface that PPGIS, as it is commonly referred to, in this sense would be of a beneficial nature to those in the community or area that is being represented. But in truth only certain groups or individuals will be able to obtain the technology and use it. Is PPGIS becoming more available to the underprivileged sector of the community? The question of "who benefits?" should always be asked, and does this harm a community or group of individuals.

The local, participatory management of urban neighborhoods usually follows on from 'claiming the territory', and has to be made compatible with national or local authority regulations on administering, managing and planning urban territory. PPGIS applied to participatory community/ neighborhood planning has been examined by, among many others. Specific attention has been given to applications such as housing issues or neighborhood revitalization. Spatial databases along with the P-mapping are used to maintain a public records GIS or community land information systems These are just a few of the uses of GIS in the community.

Examples:

Public Participation in decision making processes works not only to identify areas of common values or variability, but also as an illustrative and instructional tool. One example of effective dialogue and building trust between the community and decision makers comes from pre-planning for development in the United Kingdom. It involves using GIS and multi-criteria decision analysis (MCDA) to make a decision about wind farm siting. This method hinges upon taking all stakeholder perspectives into account to improve chances of reaching consensus. This also creates a more transparent process and adds weight to the final decision by building upon traditional methods such as public meetings and hearings, surveys, focus groups, and deliberative processes enabling participants more insights and more informed opinions on environmental issues.

Collaborative processes that consider objective and subjective inputs have the potential to efficiently address some of the conflict between development and nature as they involve a fuller justification by wind farm developers for location, scale, and design. Spatial tools such as creation of 3D view sheds offer participants new ways of assessing visual intrusion to make a more informed decision. Higgs et al. make a very telling statement when analyzing the success of this project – "the only way of accommodating people's landscape concerns is to site wind farms in places that people find more acceptable". This implies that developers recognize the validity of citizens' concerns and are willing to compromise in identifying sites where wind farms will not only be successful financially, but also successful politically and socially. This creates greater accountability and facilitates the incorporation of stakeholder values to resolve differences and gain public acceptance for vital development projects.

In another planning example, Simao et al. analyzed how to create sustainable development options with widespread community support. They determined that stakeholders need to learn likely outcomes that result from stated preferences, which can be supported through enhanced access to information and incentives to increase public participation. Through a multi-criteria spatial decision support system stakeholders were able to voice concerns and work on a compromise solution to have final outcome accepted by majority when siting wind farms. This differs from the work of Higgs et al. in that the focus was on allowing users to learn from the collaborative process, both interactively and iteratively about the nature of the problem and their own preferences for desirable characteristics of solution.

This stimulated sharing of opinions and discussion of interests behind preferences. After understanding the problem more fully, participants could discuss alternative solutions and interact with other participants to come to a compromise solution. Similar work has been done to incorporate public participation in spatial planning for transportation system development, and this method of two-way benefits is even beginning to move towards web-based mapping services to further simplify and extend the process into the community.

GEOGRAPHIC INFORMATION SCIENCE

The focal point of Geographic Information Science is the technical implementation of Geographic Information Systems. In other words, it involves the conceptual ideas for how to implement GIS.

How is GI Science Different from Geographic Information Systems?

We know all know that Geographic Information Systems looks at the "what" and "where".

For example, an electric company would store its assets in a GIS system as points, lines and polygons. The "where" is their physical geography on a map:

- Points may be towers as XY locations.
- Lines may be wires that are connected to each tower.

- Polygons may be the areas each line services.

All of these have attributes tied to them. The "what" is information about their feature:

- Towers can be made of steel, wood and other material.
- Wires can be overhead or underground.
- Service areas can have population and demographics they service.

The focal point of Geographic Information Science is the technical implementation of Geographic Information Systems. In other words, it involves the conceptual ideas for how to implement GIS.

As you know, there is unlimited potential to apply GIS in our everyday lives including these 1000 GIS applications.

GI Science Builds Better Geographic Information Systems

While Geographic Information Systems answers the "what" and "where", Geographic Information Science is concerned with the "how".

For example, GIScience conceptualizes how to store spatial information, collect data and analyze it. It encompasses all aspects of GIS such as remote sensing, surveying, mathematics, programming and geography.

Geographic Information Systems relies on the developments in GIScience for future developments. In other words, GIScience is the building block and is the foundation for all uses of a Geographic Information System.

2

Components of Geographic Information System

Some of the components of geographic information system are object based spatial data models, geospatial data models, geodata and geometric networks. The diverse applications of these components as well the major concepts related to geographic information system such as GIS mapping and geospatial intelligence have been thoroughly discussed in this chapter.

DATA MODELS FOR GIS

Data models are a set of rules and/or constructs used to describe and represent aspects of the real world in a computer. Two primary data models are available to complete this task: raster data models and vector data models.

Raster Data Models

The raster data model is widely used in applications ranging far beyond geographic information systems (GISs). Most likely, you are already very familiar with this data model if you have any experience with digital photographs. The ubiquitous JPEG, BMP, and TIFF file formats (among others) are based on the raster data model. Take a moment to view your favorite digital image. If you zoom deeply into the image, you will notice that it is composed of an array of tiny square pixels (or picture elements). Each of these uniquely colored pixels, when viewed as a whole, combines to form a coherent image.

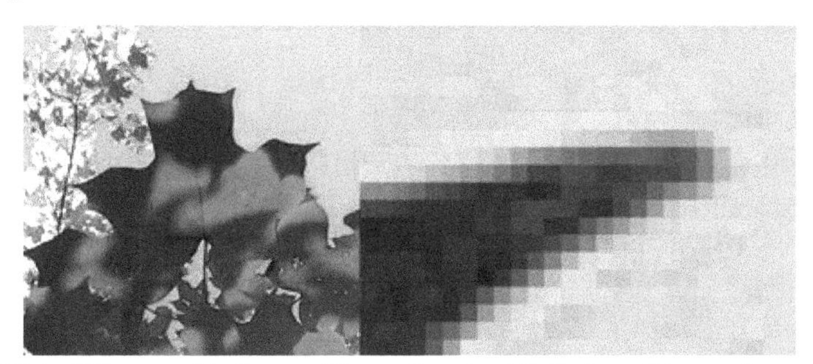

Digital Picture with Zoomed Inset Showing Pixilation of Raster Image.

Furthermore, all liquid crystal display (LCD) computer monitors are based on raster technology as they are composed of a set number of rows and columns of pixels. Notably, the foundation of this technology predates computers and digital cameras by nearly a century. The neoimpressionist artist, Georges Seurat, developed a painting technique referred to as "pointillism" in the 1880s, which similarly relies on the amassing of small, monochromatic "dots" of ink that combine to form a larger image. If you are as generous as the author, you may indeed think of your raster dataset creations as sublime works of art.

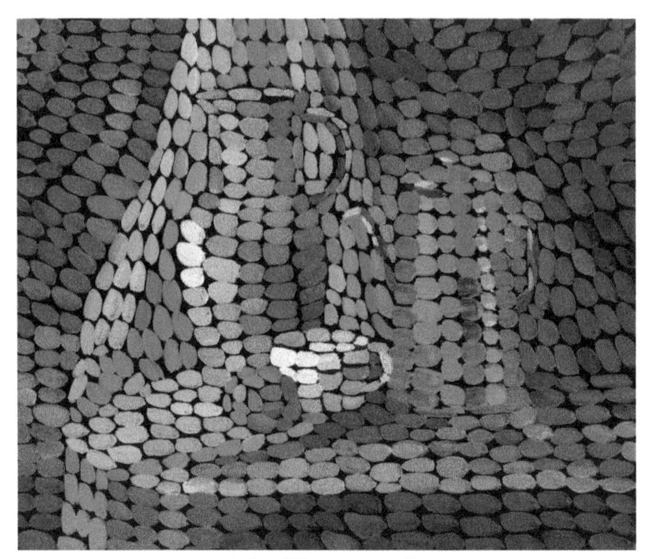

Pointillist Artwork.

The raster data model consists of rows and columns of equally sized pixels interconnected to form a planar surface. These pixels are used as building blocks for creating points, lines, areas, networks, and surfaces. illustrates how a land parcel can be converted to a raster representation). Although pixels may be triangles, hexagons, or even octagons, square pixels represent the simplest geometric form with which to work. Accordingly, the vast majority of available raster GIS data are built on the square pixel. These squares are typically reformed into rectangles of various dimensions if the data model is transformed from one projection to another.

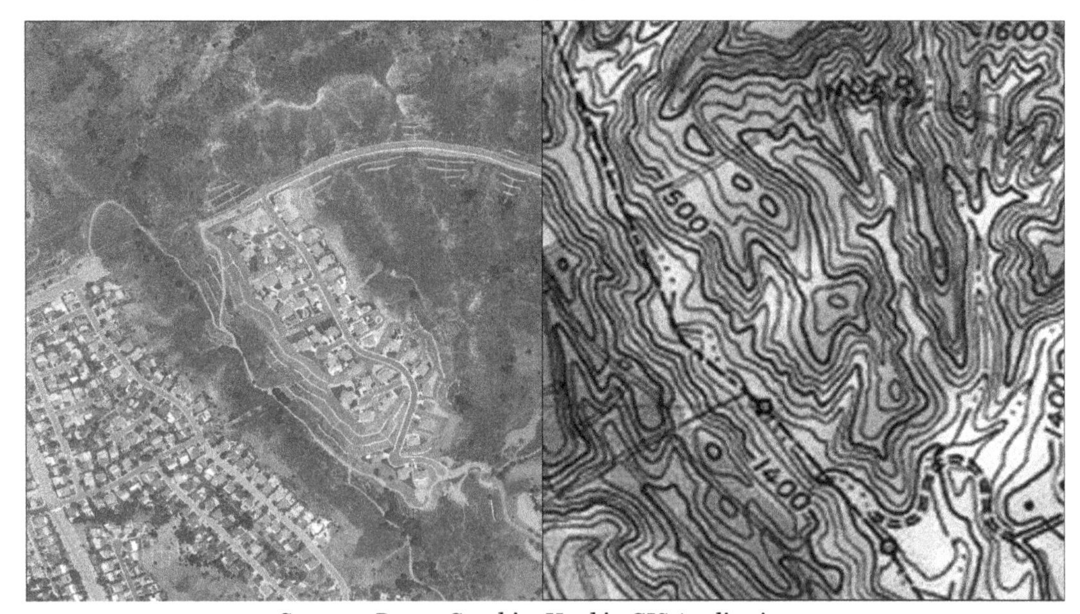

Common Raster Graphics Used in GIS Applications.

Because of the reliance on a uniform series of square pixels, the raster data model is referred to as a grid-based system. Typically, a single data value will be assigned to each grid locale. Each cell in a raster carries a single value, which represents the characteristic of the spatial phenomenon at a location denoted by its row and column. The data type for that cell value can be either integer or floating-point. Alternatively, the raster graphic can reference a database management system wherein open-ended attribute tables can be used to associate multiple data values to each pixel. The advance of computer technology has made this second methodology increasingly feasible as large datasets are no longer constrained by computer storage issues as they were previously.

The raster model will average all values within a given pixel to yield a single value. Therefore, the more area covered per pixel, the less accurate the associated data values. The area covered by each pixel determines the spatial resolution of the raster model from which it is derived. Specifically, resolution is determined by measuring one side of the square pixel. A raster model with pixels representing 10 m by 10 m (or 100 square meters) in the real world would be said to have a spatial resolution of 10 m; a raster model with pixels measuring 1 km by 1 km (1 square kilometer) in the real world would be said to have a spatial resolution of 1 km; and so forth.

Care must be taken when determining the resolution of a raster because using an overly coarse pixel resolution will cause a loss of information, whereas using overly fine pixel resolution will result in significant increases in file size and computer processing requirements during display and/or analysis. An effective pixel resolution will take both the map scale and the minimum mapping unit of the other GIS data into consideration. In the case of raster graphics with coarse spatial resolution, the data values associated with specific locations are not necessarily explicit in the raster data model. For example, if the location of telephone poles were mapped on a coarse raster graphic, it would be clear that the entire cell would not be filled by the pole. Rather, the pole would be assumed to be located somewhere within that cell (typically at the center).

Imagery employing the raster data model must exhibit several properties. First, each pixel must hold at least one value, even if that data value is zero. Furthermore, if no data are present for a given pixel, a data value placeholder must be assigned to this grid cell. Often, an arbitrary, readily identifiable value will be assigned to pixels for which there is no data value. Second, a cell can hold any alphanumeric index that represents an attribute. In the case of quantitative datasets, attribute assignation is fairly straightforward. For example, if a raster image denotes elevation, the data values for each pixel would be some indication of elevation, usually in feet or meters. In the case of qualitative datasets, data values are indices that necessarily refer to some predetermined translational rule. In the case of a land-use/land-cover raster graphic, the following rule may be applied: 1 = grassland, 2 = agricultural, 3 = disturbed, and so forth. The third property of the raster data model is that points and lines "move" to the center of the cell. As one might expect, if a 1 km resolution raster image contains a river or stream, the location of the actual waterway within the "river" pixel will be unclear. Therefore, there is a general assumption that all zero-dimensional (point) and one-dimensional (line) features will be located toward the center of the cell. As a corollary, the minimum width for any line feature must necessarily be one cell regardless of the actual width of the feature. If it is not, the feature will not be represented in the image and will therefore be assumed to be absent.

Land-Use/Land-Cover Raster Image.

Several methods exist for encoding raster data from scratch. Three of these models are as follows:

- Cell-by-cell raster encoding: This minimally intensive method encodes a raster by creating records for each cell value by row and column. This method could be thought of as a large spreadsheet wherein each cell of the spreadsheet represents a pixel in the raster image. This method is also referred to as "exhaustive enumeration."
- Run-length raster encoding: This method encodes cell values in runs of similarly valued pixels and can result in a highly compressed image file. The run-length encoding method is useful in situations where large groups of neighboring pixels have similar values (e.g., discrete datasets such as land use/land cover or habitat suitability) and is less useful where neighboring pixel values vary widely (e.g., continuous datasets such as elevation or sea-surface temperatures).
- Quad-tree raster encoding: This method divides a raster into a hierarchy of quadrants that are subdivided based on similarly valued pixels. The division of the raster stops when a quadrant is made entirely from cells of the same value. A quadrant that cannot be subdivided is called a "leaf node."

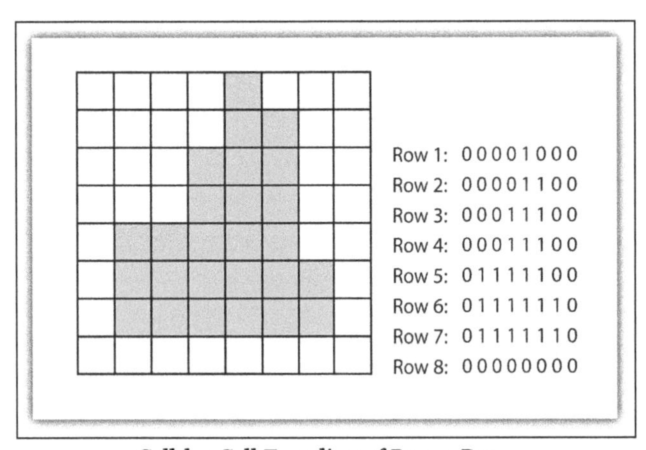

Cell-by-Cell Encoding of Raster Data.

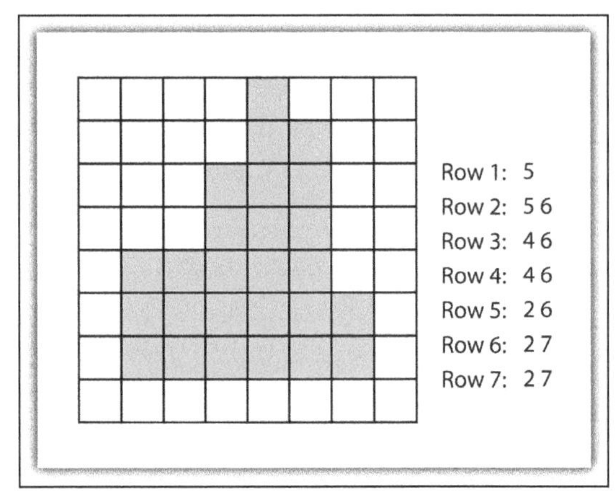

Run-Length Encoding of Raster Data.

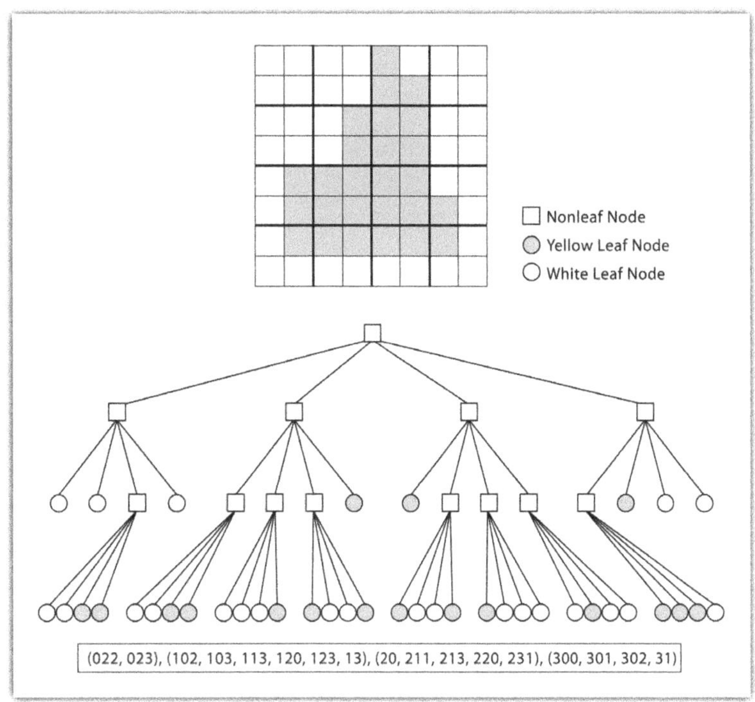

Quad-Tree Encoding of Raster Data.

Advantages/Disadvantages of the Raster Model

The use of a raster data model confers many advantages. First, the technology required to create raster graphics is inexpensive and ubiquitous. Nearly everyone currently owns some sort of raster image generator, namely a digital camera, and few cellular phones are sold today that don't include such functionality. Similarly, a plethora of satellites are constantly beaming up-to-the-minute raster graphics to scientific facilities across the globe. These graphics are often posted online for private and public use, occasionally at no cost to the user.

Additional advantages of raster graphics are the relative simplicity of the underlying data structure. Each grid location represented in the raster image correlates to a single value (or series of values if attributes tables are included). This simple data structure may also help explain why it is relatively easy to perform overlay analyses on raster data (for more on overlay analyses, This simplicity also lends itself to easy interpretation and maintenance of the graphics, relative to its vector counterpart.

Despite the advantages, there are also several disadvantages to using the raster data model. The first disadvantage is that raster files are typically very large. Particularly in the case of raster images built from the cell-by-cell encoding methodology, the sheer number of values stored for a given dataset result in potentially enormous files. Any raster file that covers a large area and has somewhat finely resolved pixels will quickly reach hundreds of megabytes in size or more. These large files are only getting larger as the quantity and quality of raster datasets continues to keep pace with quantity and quality of computer resources and raster data collectors (e.g., digital cameras, satellites).

A second disadvantage of the raster model is that the output images are less "pretty" than their vector counterparts. This is particularly noticeable when the raster images are enlarged or zoomed

Depending on how far one zooms into a raster image, the details and coherence of that image will quickly be lost amid a pixilated sea of seemingly randomly colored grid cells.

The geometric transformations that arise during map reprojection efforts can cause problems for raster graphics and represent a third disadvantage to using the raster data model. As described in, changing map projections will alter the size and shape of the original input layer and frequently result in the loss or addition of pixels. "Display of Pixel Loss and Replication in Reprojecting Raster Data from the Sinusoidal Projection." These alterations will result in the perfect square pixels of the input layer taking on some alternate rhomboidal dimensions. However, the problem is larger than a simple reformation of the square pixel. Indeed, the reprojection of a raster image dataset from one projection to another brings change to pixel values that may, in turn, significantly alter the output information.

The final disadvantage of using the raster data model is that it is not suitable for some types of spatial analyses. For example, difficulties arise when attempting to overlay and analyze multiple raster graphics produced at differing scales and pixel resolutions. Combining information from a raster image with 10 m spatial resolution with a raster image with 1 km spatial resolution will most likely produce nonsensical output information as the scales of analysis are far too disparate to result in meaningful and/or interpretable conclusions. In addition, some network and spatial analyses (i.e., determining directionality or geocoding) can be problematic to perform on raster data.

Vector Data Model

In contrast to the raster data model is the vector data model. In this model, space is not quantized into discrete grid cells like the raster model. Vector data models use points and their associated X, Y coordinate pairs to represent the vertices of spatial features, much as if they were being drawn on a map by hand. Aronoff, S. 1989. The data attributes of these features are then stored in a separate database management system. The spatial information and the attribute information for these models are linked via a simple identification number that is given to each feature in a map.

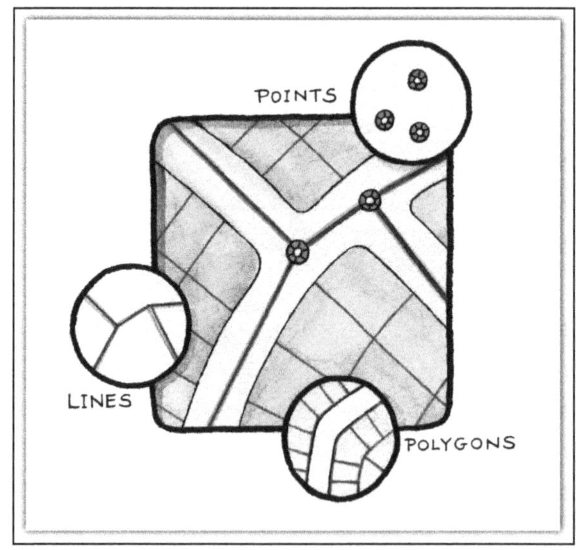

Points, Lines,and Polygons.

Three fundamental vector types exist in geographic information systems (GISs): points, lines, and polygons. Points are zero-dimensional objects that contain only a single coordinate pair. Points are typically used to model singular, discrete features such as buildings, wells, power poles, sample locations, and so forth. Points have only the property of location. Other types of point features include the node and the vertex. Specifically, a point is a stand-alone feature, while a node is a topological junction representing a common X, Y coordinate pair between intersecting lines and polygons. Vertices are defined as each bend along a line or polygon feature that is not the intersection of lines or polygons.

Points can be spatially linked to form more complex features. Lines are one-dimensional features composed of multiple, explicitly connected points. Lines are used to represent linear features such as roads, streams, faults, boundaries, and so forth. Lines have the property of length. Lines that directly connect two nodes are sometimes referred to as chains, edges, segments, or arcs.

Polygons are two-dimensional features created by multiple lines that loop back to create a "closed" feature. In the case of polygons, the first coordinate pair (point) on the first line segment is the same as the last coordinate pair on the last line segment. Polygons are used to represent features such as city boundaries, geologic formations, lakes, soil associations, vegetation communities, and so forth. Polygons have the properties of area and perimeter. Polygons are also called areas.

Vector Data Models Structures

Vector data models can be structured many different ways. We will examine two of the more common data structures here. The simplest vector data structure is called the spaghetti data model. Dangermond, J. 1982. "A Classification of Software Components Commonly Used in Geographic Information Systems." In Proceedings of the U.S.-Australia Workshop on the Design and Implementation of Computer-Based Geographic Information Systems, 70–91. Honolulu, HI. In the spaghetti model, each point, line, and polygon feature is represented as a string of X, Y coordinate pairs (or as a single X, Y coordinate pair in the case of a vector image with a single point) with no inherent structure. One could envision each line in this model to be a single strand of spaghetti that is formed into complex shapes by the addition of more and more strands of spaghetti. It is notable that in this model, any polygons that lie adjacent to each other must be made up of their own lines, or stands of spaghetti. In other words, each polygon must be uniquely defined by its own set of X, Y coordinate pairs, even if the adjacent polygons share the exact same boundary information. This creates some redundancies within the data model and therefore reduces efficiency.

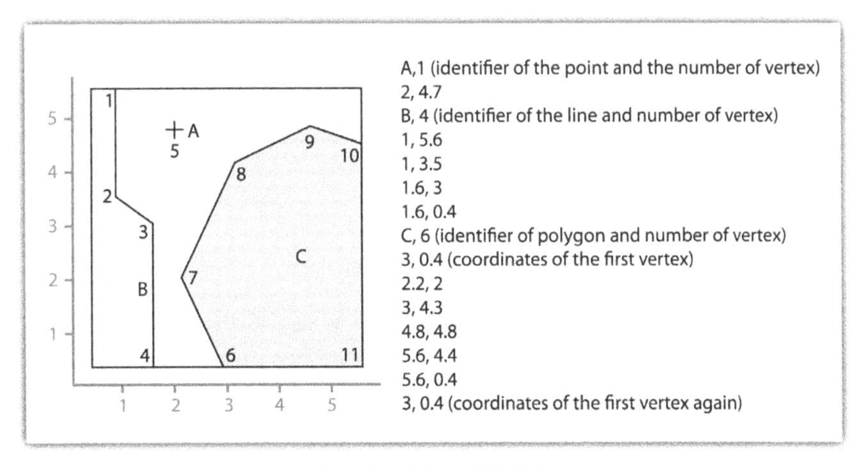

Spaghetti Data Model.

Despite the location designations associated with each line, or strand of spaghetti, spatial relationships are not explicitly encoded within the spaghetti model; rather, they are implied by their location. This results in a lack of topological information, which is problematic if the user attempts to make measurements or analysis. The computational requirements, therefore, are very steep if any advanced analytical techniques are employed on vector files structured thusly. Nevertheless, the simple structure of the spaghetti data model allows for efficient reproduction of maps and graphics as this topological information is unnecessary for plotting and printing.

In contrast to the spaghetti data model, the topological data model is characterized by the inclusion of topological information within the dataset, as the name implies. Topology is a set of rules that model the relationships between neighboring points, lines, and polygons and determines how they share geometry. For example, consider two adjacent polygons. In the spaghetti model, the shared boundary of two neighboring polygons is defined as two separate, identical lines. The inclusion of topology into the data model allows for a single line to represent this shared boundary with an explicit reference to denote which side of the line belongs with which polygon. Topology is also concerned with preserving spatial properties when the forms are bent, stretched, or placed under similar geometric transformations, which allows for more efficient projection and reprojection of map files.

Three basic topological precepts that are necessary to understand the topological data model are outlined here. First, connectivity describes the arc-node topology for the feature dataset. Nodes are more than simple points. In the topological data model, nodes are the intersection points where two or more arcs meet. In the case of arc-node topology, arcs have both a from-node (i.e., starting node) indicating where the arc begins and a to-node (i.e., ending node) indicating where the arc ends. In addition, between each node pair is a line segment, sometimes called a link, which has its own identification number and references both its from-node and to-node. In, arcs 1, 2, and 3 all intersect because they share node 11. Therefore, the computer can determine that it is possible to move along arc 1 and turn onto arc 3, while it is not possible to move from arc 1 to arc 5, as they do not share a common node.

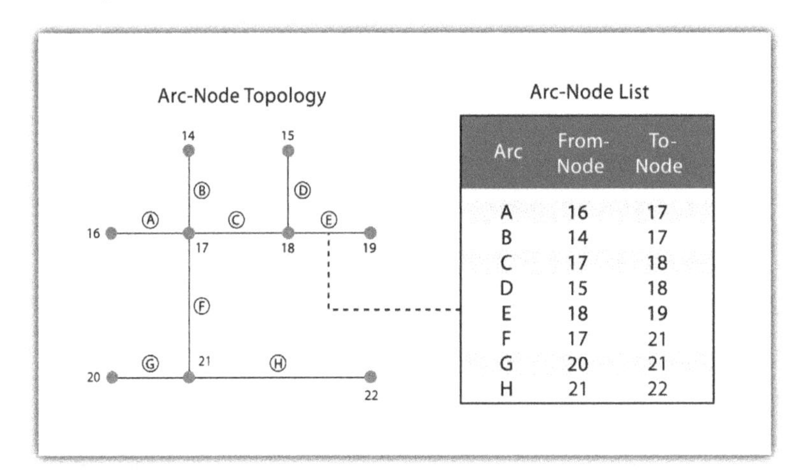

Arc	From-Node	To-Node
A	16	17
B	14	17
C	17	18
D	15	18
E	18	19
F	17	21
G	20	21
H	21	22

Arc-Node Topology.

The second basic topological precept is area definition. Area definition states that an arc that connects to surround an area defines a polygon, also called polygon-arc topology. In the case of polygon-arc topology, arcs are used to construct polygons, and each arc is stored only once. This results in a reduction in the amount of data stored and ensures that adjacent polygon boundaries do not overlap. In the, the polygon-arc topology makes it clear that polygon F is made up of arcs 8, 9, and 10.

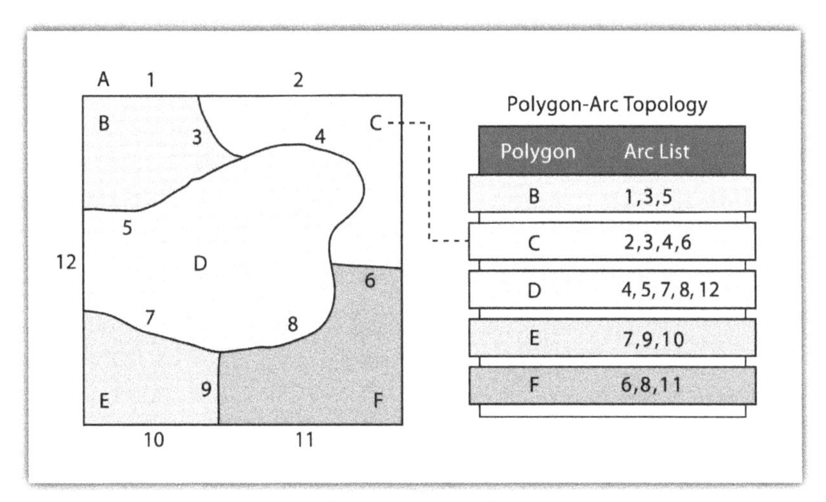

Polygon	Arc List
B	1,3,5
C	2,3,4,6
D	4, 5, 7, 8, 12
E	7,9,10
F	6,8,11

Polygon-Arc Topology.

Contiguity, the third topological precept, is based on the concept that polygons that share a boundary are deemed adjacent. Specifically, polygon topology requires that all arcs in a polygon have a direction (a from-node and a to-node), which allows adjacency information to be determined. Polygons that share an arc are deemed adjacent, or contiguous, and therefore the "left" and "right" side of each arc can be defined. This left and right polygon information is stored explicitly within the attribute information of the topological data model. The "universe polygon" is an essential component of polygon topology that represents the external area located outside of the study area. Figure "Polygon Topology" shows that arc 6 is bound on the left by polygon B and to the right by polygon C. Polygon A, the universe polygon, is to the left of arcs 1, 2, and 3.

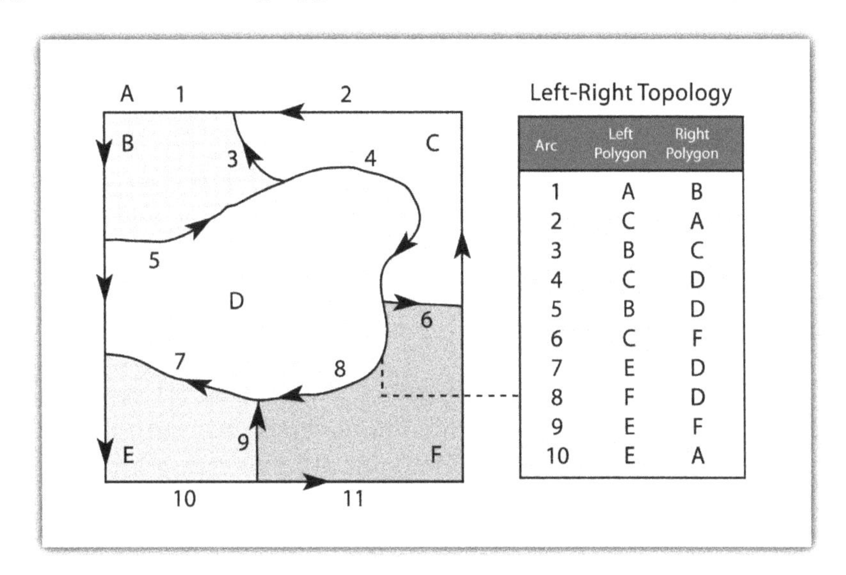

Arc	Left Polygon	Right Polygon
1	A	B
2	C	A
3	B	C
4	C	D
5	B	D
6	C	F
7	E	D
8	F	D
9	E	F
10	E	A

Polygon Topology.

Topology allows the computer to rapidly determine and analyze the spatial relationships of all its included features. In addition, topological information is important because it allows for efficient error detection within a vector dataset. In the case of polygon features, open or unclosed polygons, which occur when an arc does not completely loop back upon itself, and unlabeled polygons, which occur when an area does not contain any attribute information, violate polygon-arc topology rules.

Another topological error found with polygon features is the sliver. Slivers occur when the shared boundary of two polygons do not meet exactly.

In the case of line features, topological errors occur when two lines do not meet perfectly at a node. This error is called an "undershoot" when the lines do not extend far enough to meet each other and an "overshoot" when the line extends beyond the feature it should connect to. The result of overshoots and undershoots is a "dangling node" at the end of the line. Dangling nodes aren't always an error, however, as they occur in the case of dead-end streets on a road map.

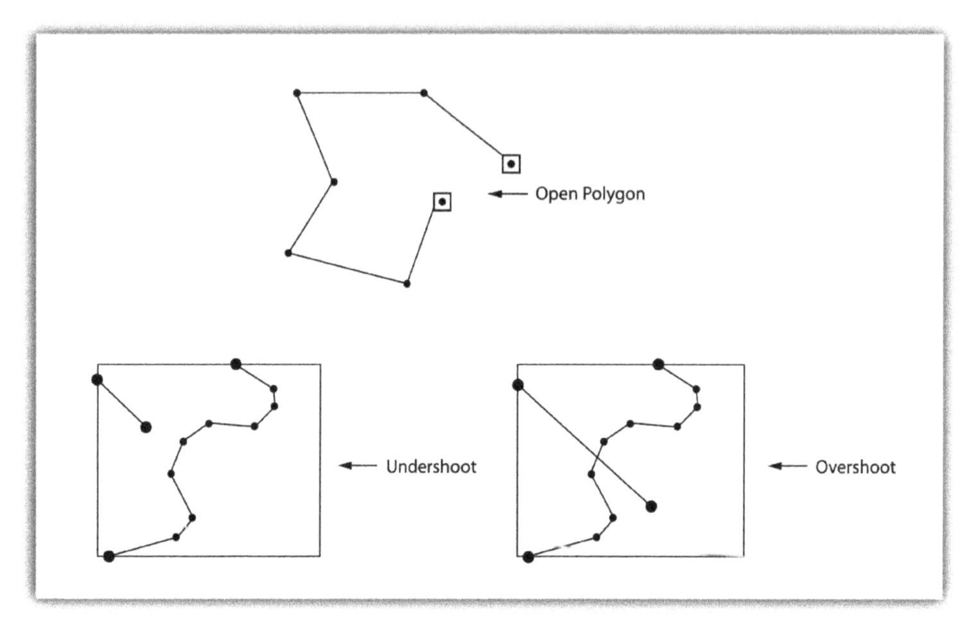

Common Topological Errors.

Many types of spatial analysis require the degree of organization offered by topologically explicit data models. In particular, network analysis (e.g., finding the best route from one location to another) and measurement (e.g., finding the length of a river segment) relies heavily on the concept of to- and from-nodes and uses this information, along with attribute information, to calculate distances, shortest routes, quickest routes, and so forth. Topology also allows for sophisticated neighborhood analysis such as determining adjacency, clustering, nearest neighbors, and so forth.

Now that the basics of the concepts of topology have been outlined, we can begin to better understand the topological data model. In this model, the node acts as more than just a simple point along a line or polygon. The node represents the point of intersection for two or more arcs. Arcs may or may not be looped into polygons. Regardless, all nodes, arcs, and polygons are individually numbered. This numbering allows for quick and easy reference within the data model.

Advantages/Disadvantages of the Vector Model

In comparison with the raster data model, vector data models tend to be better representations of reality due to the accuracy and precision of points, lines, and polygons over the regularly spaced grid cells of the raster model. This results in vector data tending to be more aesthetically pleasing than raster data.

Vector data also provides an increased ability to alter the scale of observation and analysis. As each coordinate pair associated with a point, line, and polygon represents an infinitesimally exact location (albeit limited by the number of significant digits and/or data acquisition methodologies), zooming deep into a vector image does not change the view of a vector graphic in the way that it does a raster graphic.

Vector data tend to be more compact in data structure, so file sizes are typically much smaller than their raster counterparts. Although the ability of modern computers has minimized the importance of maintaining small file sizes, vector data often require a fraction the computer storage space when compared to raster data.

The final advantage of vector data is that topology is inherent in the vector model. This topological information results in simplified spatial analysis (e.g., error detection, network analysis, proximity analysis, and spatial transformation) when using a vector model.

Alternatively, there are two primary disadvantages of the vector data model. First, the data structure tends to be much more complex than the simple raster data model. As the location of each vertex must be stored explicitly in the model, there are no shortcuts for storing data like there are for raster models (e.g., the run-length and quad-tree encoding methodologies).

Second, the implementation of spatial analysis can also be relatively complicated due to minor differences in accuracy and precision between the input datasets. Similarly, the algorithms for manipulating and analyzing vector data are complex and can lead to intensive processing requirements, particularly when dealing with large datasets.

Satellite Imagery and Aerial Photography

A wide variety of satellite imagery and aerial photography is available for use in geographic information systems (GISs). Although these products are basically raster graphics, they are substantively different in their usage within a GIS. Satellite imagery and aerial photography provide important contextual information for a GIS and are often used to conduct heads-up digitizing whereby features from the image are converted into vector datasets.

Satellite Imagery

Remotely sensed satellite imagery is becoming increasingly common as satellites equipped with technologically advanced sensors are continually being sent into space by public agencies and private companies around the globe. Satellites are used for applications such as military and civilian earth observation, communication, navigation, weather, research, and more. Currently, more than 3,000 satellites have been sent to space, with over 2,500 of them originating from Russia and the United States. These satellites maintain different altitudes, inclinations, eccentricities, synchronies, and orbital centers, allowing them to image a wide variety of surface features and processes.

Satellites can be active or passive. Active satellites make use of remote sensors that detect reflected responses from objects that are irradiated from artificially generated energy sources. For example, active sensors such as radars emit radio waves, laser sensors emit light waves, and sonar sensors emit sound waves. In all cases, the sensor emits the signal and then calculates the time it takes for the returned signal to "bounce" back from some remote feature. Knowing the speed of the emitted signal, the time delay from the original emission to the return can be used to calculate the distance to the feature.

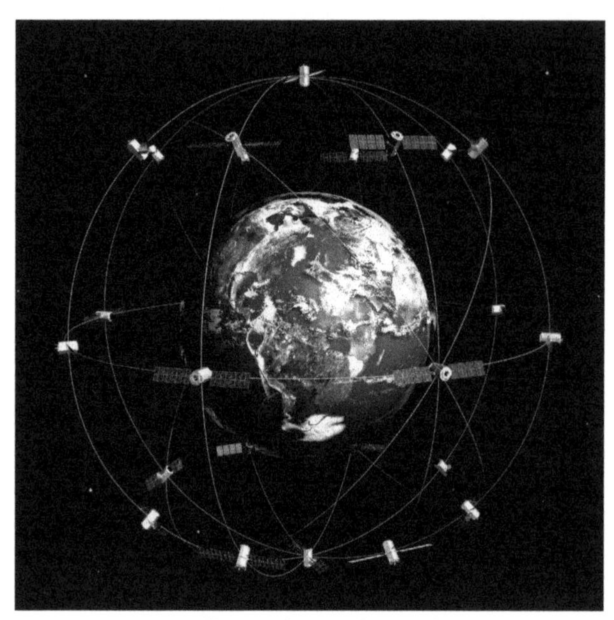

Satellites Orbiting the Earth.

Passive satellites, alternatively, make use of sensors that detect the reflected or emitted electromagnetic radiation from natural sources. This natural source is typically the energy from the sun, but other sources can be imaged as well, such as magnetism and geothermal activity. Using an example we've all experienced, taking a picture with a flash-enabled camera would be active remote sensing, while using a camera without a flash (i.e., relying on ambient light to illuminate the scene) would be passive remote sensing.

The quality and quantity of satellite imagery is largely determined by their resolution. There are four types of resolution that characterize any particular remote sensor. The spatial resolution of a satellite image, as described previously in the raster data model section, is a direct representation of the ground coverage for each pixel shown in the image. If a satellite produces imagery with a 10 m resolution, the corresponding ground coverage for each of those pixels is 10 m by 10 m, or 100 square meters on the ground. Spatial resolution is determined by the sensors' instantaneous field of view (IFOV). The IFOV is essentially the ground area through which the sensor is receiving the electromagnetic radiation signal and is determined by height and angle of the imaging platform.

Spectral resolution denotes the ability of the sensor to resolve wavelength intervals, also called bands, within the electromagnetic spectrum. The spectral resolution is determined by the interval size of the wavelengths and the number of intervals being scanned. Multispectral and hyperspectral sensors are those sensors that can resolve a multitude of wavelengths intervals within the spectrum. For example, the IKONOS satellite resolves images for bands at the blue (445–516 nm), green (506–95 nm), red (632–98 nm), and near-infrared (757–853 nm) wavelength intervals on its 4-meter multispectral sensor.

Temporal resolution is the amount of time between each image collection period and is determined by the repeat cycle of the satellite's orbit. Temporal resolution can be thought of as true-nadir or off-nadir. Areas considered true-nadir are those located directly beneath the sensor while off-nadir areas are those that are imaged obliquely. In the case of the IKONOS satellite, the temporal resolution is 3 to 5 days for off-nadir imaging and 144 days for true-nadir imaging.

The fourth and final type of resolution, radiometric resolution, refers to the sensitivity of the sensor to variations in brightness and specifically denotes the number of grayscale levels that can be imaged by the sensor. Typically, the available radiometric values for a sensor are 8-bit (yielding values that range from 0–255 as 256 unique values or as 2^8 values); 11-bit (0–2,047); 12-bit (0–4,095); or 16-bit (0–63,535). Landsat-7, for example, maintains 8-bit resolution for its bands and can therefore record values for each pixel that range from 0 to 255.

Because of the technical constraints associated with satellite remote sensing systems, there is a trade-off between these different types of resolution. Improving one type of resolution often necessitates a reduction in one of the other types of resolution. For example, an increase in spatial resolution is typically associated with a decrease in spectral resolution, and vice versa. Similarly, geostationary satellites (those that circle the earth proximal to the equator once each day) yield high temporal resolution but low spatial resolution, while sun-synchronous satellites (those that synchronize a near-polar orbit of the sensor with the sun's illumination) yield low temporal resolution while providing high spatial resolution. Although technological advances can generally improve the various resolutions of an image, care must always be taken to ensure that the imagery you have chosen is adequate to the represent or model the geospatial features that are most important to your study.

Aerial Photography

Aerial photography, like satellite imagery, represents a vast source of information for use in any GIS. Platforms for the hardware used to take aerial photographs include airplanes, helicopters, balloons, rockets, and so forth. While aerial photography connotes images taken of the visible spectrum, sensors to measure bands within the nonvisible spectrum (e.g., ultraviolet, infrared, near-infrared) can also be fixed to aerial sources. Similarly, aerial photography can be active or passive and can be taken from vertical or oblique angles. Care must be taken with aerial photographs as the sensors used to take the images are similar to cameras in their use of lenses. These lenses add a curvature to the images, which becomes more pronounced as one moves away from the center of the photo.

Curvature Error Due to Lenticular Properties of Camera.

Another source of potential error in an aerial photograph is relief displacement. This error arises from the three-dimensional aspect of terrain features and is seen as apparent leaning away of vertical objects from the center point of an aerial photograph. To imagine this type of error, consider that a smokestack would look like a doughnut if the viewing camera was directly above the feature. However, if this same smokestack was observed near the edge of the camera's view, one could observe the sides of the smokestack. This error is frequently seen with trees and multistory buildings and worsens with increasingly taller features.

Orthophotos are vertical photographs that have been geometrically "corrected" to remove the curvature and terrain-induced error from images. The most common orthophoto product is the digital ortho quarter quadrangle (DOQQ). DOQQs are available through the US Geological Survey (USGS), who began producing these images from their library of 1:40,000-scale National Aerial Photography Program photos. These images can be obtained in either grayscale or color with 1-meter spatial resolution and 8-bit radiometric resolution. As the name suggests, these images cover a quarter of a USGS 7.5 minute quadrangle, which equals an approximately 25 square mile area. Included with these photos is an additional 50 to 300-meter edge around the photo that allows users to mosaic many DOQQs into a single, continuous image. These DOQQs are ideal for use in a GIS as background display information, for data editing, and for heads-up digitizing.

OBJECT BASED SPATIAL DATA MODEL

An object-based spatial database is a spatial database that stores the location as objects. The object-based spatial model treats the world as surface littered with recognizable objects (e.g. cities, rivers), which exist independent of their locations.

Objects can be simple as polygons and lines, or be more complex to represent cities.

While a field-based data model sees the world as a continuous surface over which features (e.g. elevation) vary, using an object-based spatial database, it is easier to store additional attributes with the objects, such as direction, speed, etc. Using these attributes can make it easier to answer queries like "find all tanks whose speed is 10 km and oriented to north". Or "find all enemy tanks in a certain region".

Storing attributes with objects can provide better result presentation and improved manipulation capabilities in a more efficient way. In a field-based data model, this information is usually stored at different layers and it is harder to extract different information from various layers. This data model can be applied above the ER as in GERM model and GISER.

S. Shekhar introduces direction as a spatial object and presents a solution to object-direction-based queries.

Data Model Representation

The most common representations for the data model follow.

PostGIS

An open-source software program that adds support for geographic objects to the PostgreSQL object

-relational database. PostGIS follows the Simple Features for SQL specification from the Open Geospatial Consortium.

OMT-G

Provides a UML representation for geographic applications, it can represent the concept of field, object and provides a way to differentiate between spatial relation and simple association.

Entity Relationship

GraphDB

Represents a framework of objects as classes that are partitioned into three kinds of classes: simple classes, link classes, and path classes. Objects of a simple class are on the one hand just like objects in other models. They have an object type and an object identity and can have attributes whose values are either of a data type (e.g. integer, string) or of an object type (that is, an attribute may contain a reference to another object). So the structure of an object is basically that of a tuple or record. On the other hand, objects of a simple class are nodes of the database graph – the whole database can also be viewed as a single graph. Objects of a link class are like objects of a simple class but additionally contain two distinguished references to source and target objects (belonging to simple classes), which makes them edges of the database graph. Finally, an object of a path class is like an object of a simple class, but contains additionally a list of references to node and edge objects which form a path over the database graph.

GEIS

Represent a data model to store geographic information on top of EER model, GEIS define the input data model and provide the following for data model Geometry. In the GISER model, geometry is an entity that is related to a spatial object by the relationship determines shape of. Additional entities represent the primitives such as points, lines, and polygons as proposed in related models. Topology. Topology is a property belonging to a spatial object and that property remains unaltered even when the object deforms. An example is a road network. The two nodes in the network thus remain connected even if the path between the nodes is changed by road construction. In order to represent the topology, the basic primitives such as networks (i.e., graphs) and partitions are provided. Additional primitives can be added on lines of the Worboy model, This system support representation for stored data.

GeoOOA

Oracle Spatial

Oracle spatial is a component of enterprise Oracle 10g and provides support to stores object such as road on top of the current implentend construction but it used network data model to store geographic data as nodes and links (a graph representation) with each node or links it has a set of attributes. For example, a route object can be added to the database.

GRASS GIS

It supports raster and some set of vector representation.

GEOSPATIAL DATA MODEL AND ITS TYPES

Spatial data are what drive a GIS. Every functionality that makes a GIS separate from another analytical environment is rooted in the spatially explicit nature of the data.

Spatial data are often referred to as layers, coverages, or layers. We will use the term layers from this point on, since this is the recognized term used in ArcGIS. Layers represent, in a special digital storage format, features on, above, or below the surface of the earth. Depending on the type of features they represent, and the purpose to which the data will be applied, layers will be one of 2 major types:

- Vector data represent features as discrete points, lines, and polygons.
- Raster data represent the landscape as a rectangular matrix of square cells.

Depending on the type of problem that needs to be solved, the type of maps that need to be made, and the data source, either raster or vector, or a combination of the two can be used. Each data model has strengths and weaknesses in terms of functionality and representation. As you get more experience with GIS, you will be able to determine which data type to use for a particular application.

Data Structures

The 2 basic data structures in any fully-functional GIS are:

- Vector:
    - ArcInfo Coverages
    - ArcGIS Shape Files
    - CAD (AutoCAD DXF & DWG, or MicroStation DGN files)
    - ASCII coordinate data
- Raster:
    - ArcInfo Grids
    - Images
    - Digital Elevation Models (DEMs)
    - generic raster datasets

Vector data are composed of:

- Points
- Lines
- Polygons

These features are the basic features in a vector-based GIS, such as ArcGIS 9. The basic spatial data model is known as "arc-node topology." One of the strengths of the vector data model is that it can be used to render geographic features with great precision.. However, this comes at the cost of greater complexity in data structures, which sometimes translates to slow processing speed. Most of the features you see on printed maps are represented with vector data.

Points represent discrete locations on the ground. Either these are true points, such as the point marked by a USGS brass cap, such as a section corner, or they may be virtual points, based on the scale of representation. For example, a city's location on a driving map of the United States is represented by a point, even though in reality a city has area. As the map's scale increases, the city will soon appear as a polygon. Beyond a certain scale of zoom (i.e., when the map's extent is completely within the city), there will be no representation of the city at all; it will simply be the background of the map.

Here is a view of the Puget Sound area with airports. The airports are stored as points within the GIS.

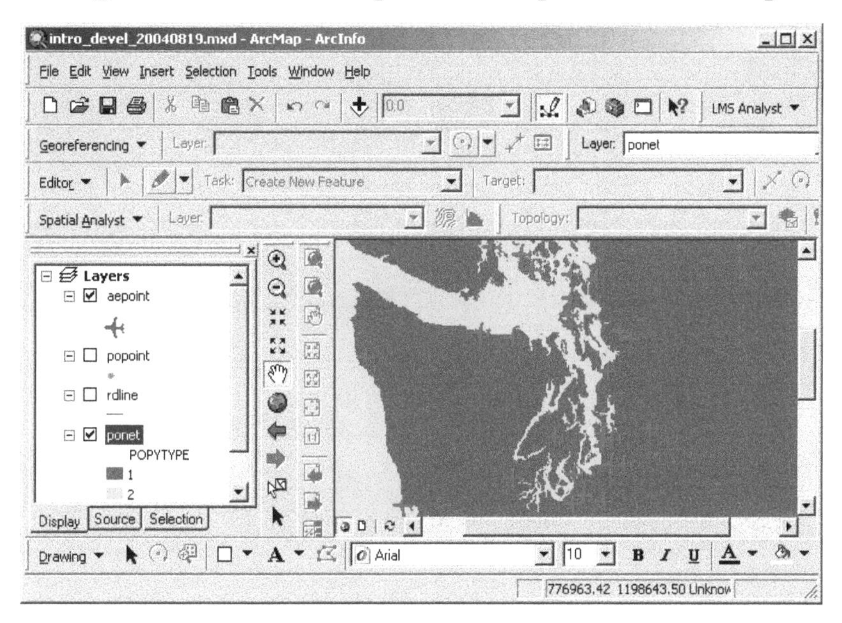

Each layer is a combination of the coordinate (vector) data, and an attribute table containing a record for each vector feature. The records hold attributes for the feature, such as city name, sampling point number, or radio tower frequency. In this example, airports represented as points, and are associated with their name as well as with other codes in the point layer attribute table.

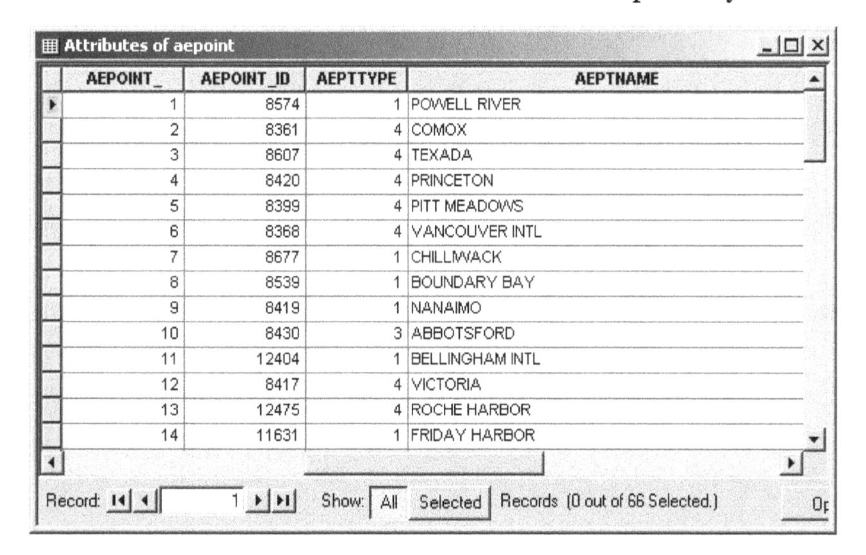

Attributes of aepoint

AEPOINT_	AEPOINT_ID	AEPTTYPE	AEPTNAME
1	8574	1	POWELL RIVER
2	8361	4	COMOX
3	8607	4	TEXADA
4	8420	4	PRINCETON
5	8399	4	PITT MEADOWS
6	8368	4	VANCOUVER INTL
7	8677	1	CHILLIWACK
8	8539	1	BOUNDARY BAY
9	8419	1	NANAIMO
10	8430	3	ABBOTSFORD
11	12404	1	BELLINGHAM INTL
12	8417	4	VICTORIA
13	12475	4	ROCHE HARBOR
14	11631	1	FRIDAY HARBOR

Record: 1 Show: All Selected Records (0 out of 66 Selected.)

Lines represent linear features, such as rivers, roads and transmission cables. Here are major roads in the Puget Sound region, along with line attributes. In ArcGIS, lines are also known as "arcs," hence the name "ArcGIS." Each line is composed of a number of different coordinates, which make up the shape of the line, as well as the tabular record for the line vector feature.

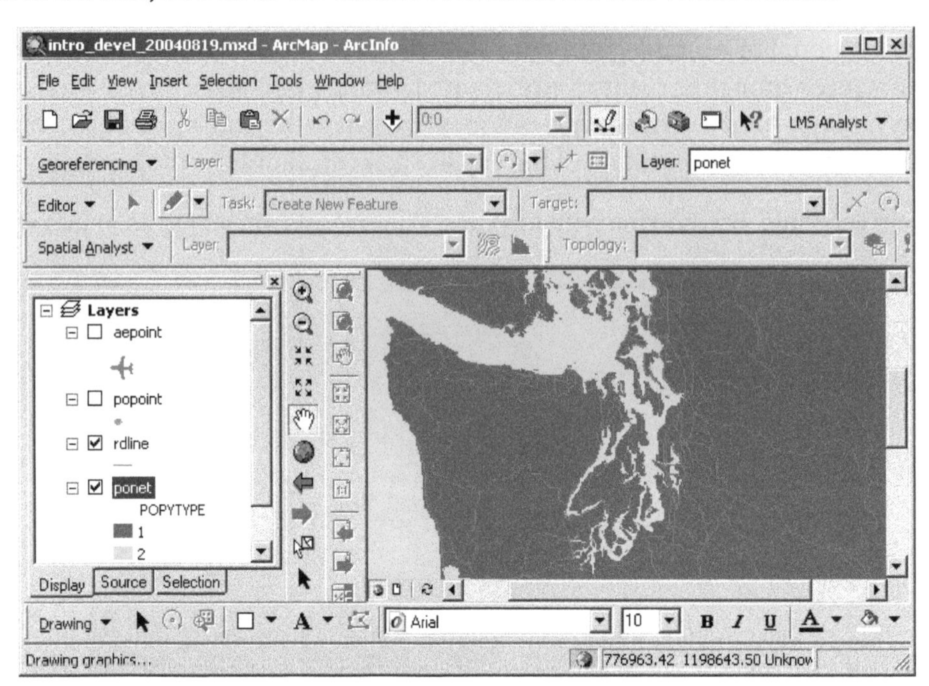

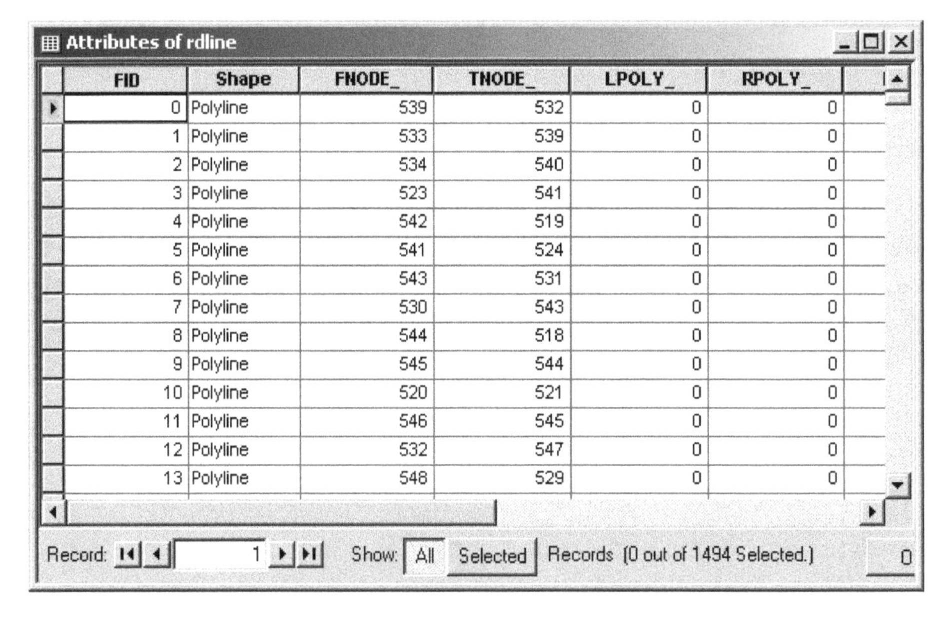

Attributes of rdline

FID	Shape	FNODE_	TNODE_	LPOLY_	RPOLY_
0	Polyline	539	532	0	0
1	Polyline	533	539	0	0
2	Polyline	534	540	0	0
3	Polyline	523	541	0	0
4	Polyline	542	519	0	0
5	Polyline	541	524	0	0
6	Polyline	543	531	0	0
7	Polyline	530	543	0	0
8	Polyline	544	518	0	0
9	Polyline	545	544	0	0
10	Polyline	520	521	0	0
11	Polyline	546	545	0	0
12	Polyline	532	547	0	0
13	Polyline	548	529	0	0

Record: 1 Show: All Selected Records (0 out of 1494 Selected.) 0

Arcs are composed of nodes and vertices. Arcs begin and end at nodes, and may have 0 or more vertices between the nodes. The vertices define the shape of the arc along its length. Arcs which connect to each other will share a common node.

Sometimes several arcs are connected and share common attributes. These arcs may appear to be a continuous line, but in reality, the "line" is composed of multiple features, each with its own record.

The arc-node topology data model is central to many ArcGIS vector operations. Arcs are represented with starting and ending nodes, which imparts directionality to the arcs. In the image below, arc #1 starts at node #2 and ends at node#1, passing through several vertices along its way. Each of the nodes and vertices is stored with coordinate values representing real-world locations in a real-world coordinate system (e.g., longitude/latitude angles, State Plane feet, or UTM meters). These coordinate values represent locations you could locate using the numeric values printed on the graticule on the edges of a paper map. The storage of real-world coordinate values for features stored in the GIS is known as georeferencing. If features stored in the GIS are referenced to real world locations, the features are said to be geoferenced.

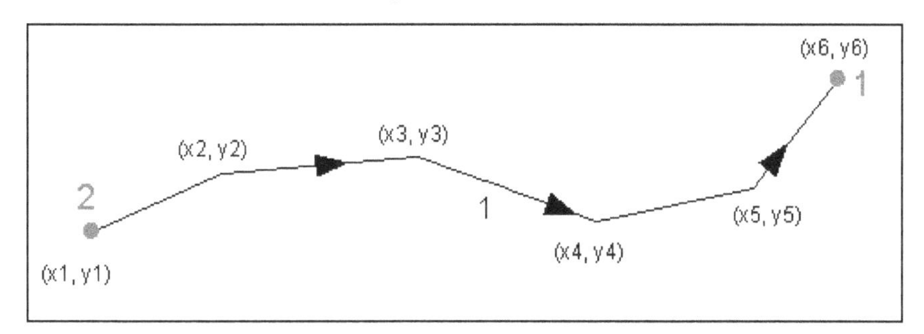

Note that each node and vertex has particular X and Y coordinate values. These X and Y coordinates form the basis of the location of features in coordinate (map) space. All vector features (points, lines, and polygons) are composed of locations defined by particular coordinate values. In the ArcGIS shapefile data model, the coordinate values on points, nodes, and vertices are stored within the dataset as "hidden" values on a feature-by-feature basis.

Polygons form bounded areas. In the point and line datasets shown above, the land masses, islands, and water features are represented as polygons. Polygons are formed by bounding arcs, which keep track of the location of each polygon, as shown in this image:

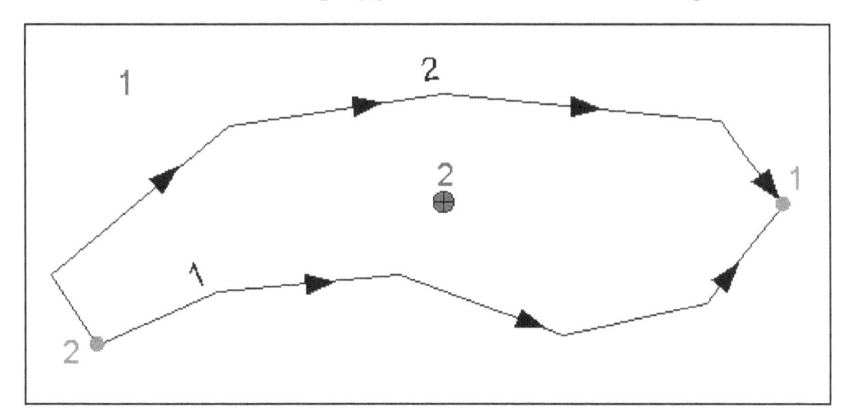

Polygon #1 is the "universe polygon," representing the outside of all other polygons. Polygon #2 (which is also denoted by labelpoint 2) is bounded by arcs 1 and 2. (Although all of these features have coordinates, they are not displayed here as they were in the previous image.)

Now consider arc #2. It starts at node #2, and ends at node #1. Following its direction, on its left-hand side is polygon #1 (which needs no label as the "universe" polygon), and on its right side is polygon #2 (denoted by labelpoint 2).

Continuing with this pattern, look at the same dataset with one added polygon:

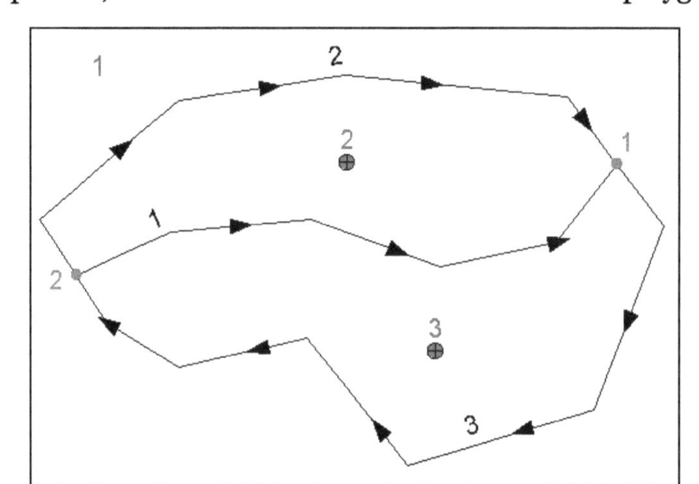

You can see now that managing vector datasets is complex. Each node, label point, vertex, line, must be stored with explicit coordinates, and the software also needs to keep track of the spatial relationships of each feature as well as the relational link with the tabular data. Imagine a dataset composed of several hundred thousand polygons; this dataset will take a lot of computing resources to manage.

Tics manage the georeferencing of almost all ArcInfo vector datasets. While tics are *not* a part of the native ArcGIS shapefile data specification, they are central to the ArcInfo vector data model, and because many of our data sources are ArcInfo coverages, this feature deserves to be mentioned. Tics are registered to ground coordinates, and all other data in the coverage are in turn referenced to the tics (which in turn makes the features referenced to ground coordinates).

Below is an image that shows all of the major features of an ArcInfo polygon dataset (technically known as an "ArcInfo coverage"). Tics are shown in cyan; arcs in black; nodes in green, and polygon label points in red.

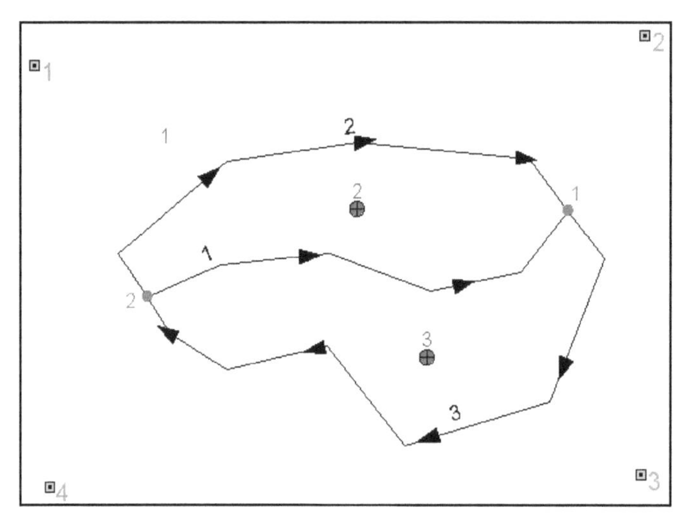

Although these spatial relationships are complex, you, as the user, do not need to keep track of these relationships; the software keeps track of all this for you. While a human can easily see the relationships among all of these features, the software lacks intelligence, and must therefore explicitly store all of the locations and relationships of features in digital files.

ArcGIS can automatically store line length and polygon area and perimeter in the units of measure defined by the user. These units can be feet, meters, or whatever units the user needs.

In ArcGIS there are 3 major types of vector data source files:

- ESRI geodatabases,
- ESRI shape files,
- ArcInfo coverages.

ESRI geodatabases are a relatively new format. Geodatabases are databases stored in Microsoft Access (for the "personal" geodatabase), as a special collection of files (for the "file-based" geodatabase), or higher-end applications (e.g. SQL Server, Oracle, Informix). A geodatabase stores all features and related tables, as well as other files, within a single or distributed database format. There are several advantages to using geodatabases rather than other storage formats: portability, integrity, validation of allowed data values, storage of data relationships as part of the data structure, topological rules, etc. At ArcGIS, it is possible to create geoprocessing models for complex analyses, as well as toolboxes containing custom tools, and have these stored in a geodatabase. In terms of the basic representation of spatial features (points, lines, and polygons), and their spatial referencing, the geodatabase functions the same as the other formats. To find out more about geodatabases, read the ArcGIS help topics. ESRI also publishes a textbook on data modeling that is focused around the geodatabase.

ESRI Shape Files are used mainly in ArcView 3.x and ArcGIS, although supported in other software as well. Because of the simple data and file structure, shapefiles draw very quickly in ArcGIS 10. Shapefiles can be fully managed (created, edited, and deleted) within ArcGIS 10's environment. Shapefile data files can also be managed using operating system tools, such as the Windows Explorer. The shapefile standard is public, so any software can be made to read or write shapefiles.

A single shapefile represents features that are either point, line, or polygon in spatial data type. If you create a shapefile, you need to choose what feature type you want at the time of creation.

Spatial referencing of shapefiles is enforced by maintaining explicit X and Y coordinates for each point or vertex in the layer. Typically, this is done at the time of data creation, where a new dataset is drawn in reference to existing datasets that are already georeferenced.

For each shapefile there exist at least 3 files, the shape data (stored in the .shp file), an associated dBASE (relational database) table (stored in the .dbf file), and a spatial index (stored in the .shx file). For each feature within the shapefile, there is an associated record within the attribute table.

ArcInfo Coverages are a basic implementation of the vector arc-node topological model as shown in the cartoon-like images above. Like shapefiles, coverages also have associated database tables, with a one-to-one feature-to-record relationship.

As you can see in the schematic for the ArcInfo polygon coverage, the coverage can be a multi-feature, or "polymorphic" dataset, composed of polygons, arcs, nodes, label points, and tics. This is due to the original way features were modeled in some of the first GIS software applications. It is a complex and sometimes unwieldy way to store data, but because many systems still use ArcInfo as the main GIS software, the coverage will be around for quite some time.

Because of the complex structure of some coverage data, coverages can take a long time to draw. Also, due to intricacies of the file system storage model, coverages cannot be managed fully (i.e., created, edited, deleted) without the use of ArcInfo software. Operating system tools for file management (e.g., the Windows Explorer) cannot manage ArcInfo datasets without corrupting them. For these reasons, ArcInfo coverages are often used as sources in ArcGIS, but they are frequently converted to the shapefile format for other uses.

Spatial referencing of ArcInfo coverages is enforced by tics, and all other features within the coverage are spatially referenced relative to the tics.

Most of the datasets we will be using throughout the term are ArcInfo coverages. Within ArcGIS, there are only minor differences between the functionality of the shapefile and the coverage.

The ArcInfo dataset file structure causes all sorts of problems for ArcGIS 10 users.

For polygon features, shapefiles and coverages have very different spatial data structures. The greatest differences can be shown comparing how polygons are stored. While coverages use the standard arc-node topology data structure, in which adjacent polygons share common bounding arcs, shapefiles store each feature as a separate object.

Here is a close up view of a few adjacent coverage polygons. When one bounding arc is moved, both polygons are affected.

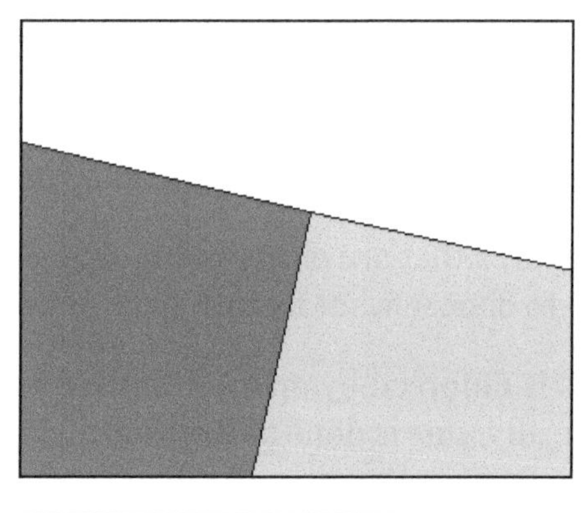

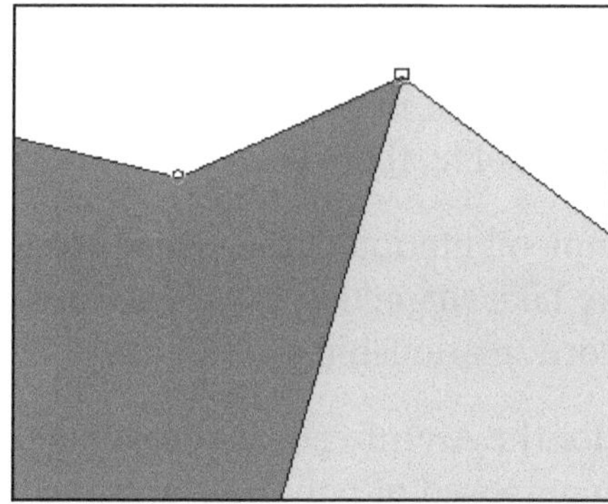

Compare this to a shapefile, in which different polygons can be moved without affecting adjacent polygons:

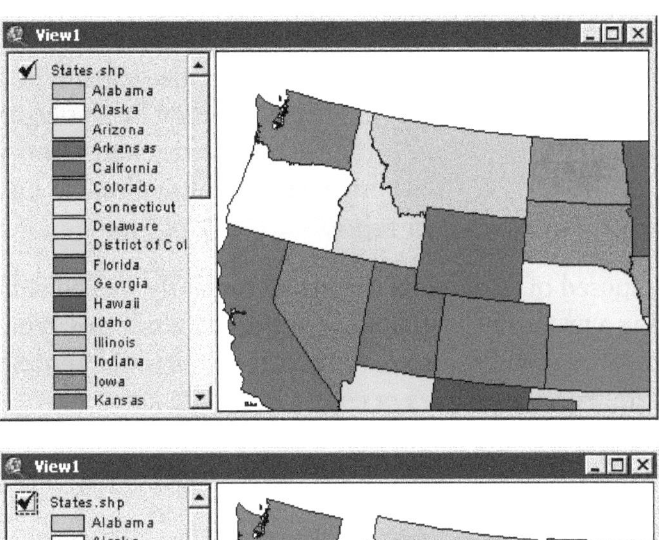

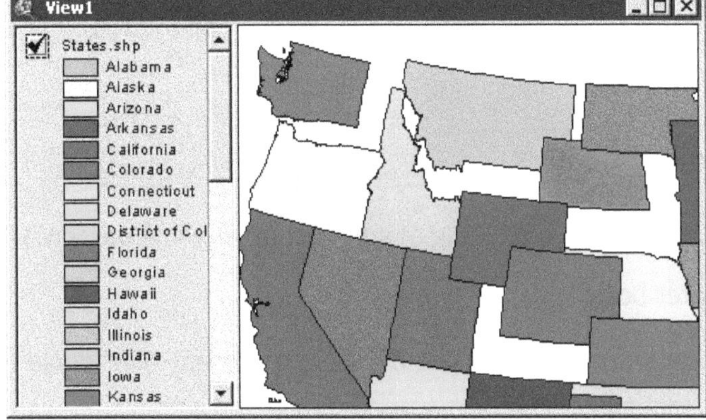

The geodatabase can function in either of the ways shown above. While polygonal features in a geodatabase are stored as individual "rings", it is possible to define topological relationships so that adjacent features are not able to be edited independently.

ASCII Coordinate Data files may also be used in ArcGIS. Point layers can be created from files containing single records for individual points. The source files which single records for individual points, where each record contains X and Y coordinates, as well as any other optional attribute data. In this example, there is a different record for each point representing a populated place in the dataset.

table2.txt - Notepad

File Edit Search Help

```
"Popyreg","Popycoun","Popyadmin","X-coord","Y-coord"
N,CA,BRITISH COLUMBIA,598110.00000,5485050.00000
5,OO,,485374.06250,5373037.00000
N,CA,BRITISH COLUMBIA,362882.12500,5538235.75000
N,CA,BRITISH COLUMBIA,403992.84375,5525419.25000
5,OO,,379086.18788,5538058.75000
N,CA,BRITISH COLUMBIA,455909.82813,5504394.50000
N,CA,BRITISH COLUMBIA,478188.09781,5534007.50000
N,CA,BRITISH COLUMBIA,370181.15625,5533627.50000
N,CA,BRITISH COLUMBIA,381337.18750,5524611.75000
N,CA,BRITISH COLUMBIA,416700.35938,5434016.25000
N,CA,BRITISH COLUMBIA,400327.21875,5499948.00000
N,CA,BRITISH COLUMBIA,429264.20313,5515108.75000
```

The coordinates can be represented as points on a map.

Vector Data Scale Dependency

For all vector datasets, you should always consider the scale dependency of spatial data. When should an airport be represented as a point, and when should it be a polygon? If you are measuring the distance from major cities to their airports, then the cities and airports would be best represented as points. However, if you are planning wetland mitigation for an addition to an airport, then the airport boundary would be better represented as a polygon.

Raster datasets are composed of rectangular arrays of regularly spaced square grid cells. Each cell has a value, representing a property or attribute of interest. While any type of geographic data can be stored in raster format, raster datasets are especially suited to the representation of continuous, rather than discrete, data. Some examples of continuous data are:

- Oil depth across an open-water oil spill
- Soil pH
- Reflectance in a certain band in the electromagnetic spectrum
- Elevation
- Landform aspect (compass bearing of steepest downward descent)
- Salinity of a water body

Here is a diagrammatic model of how raster datasets represent real-world features:

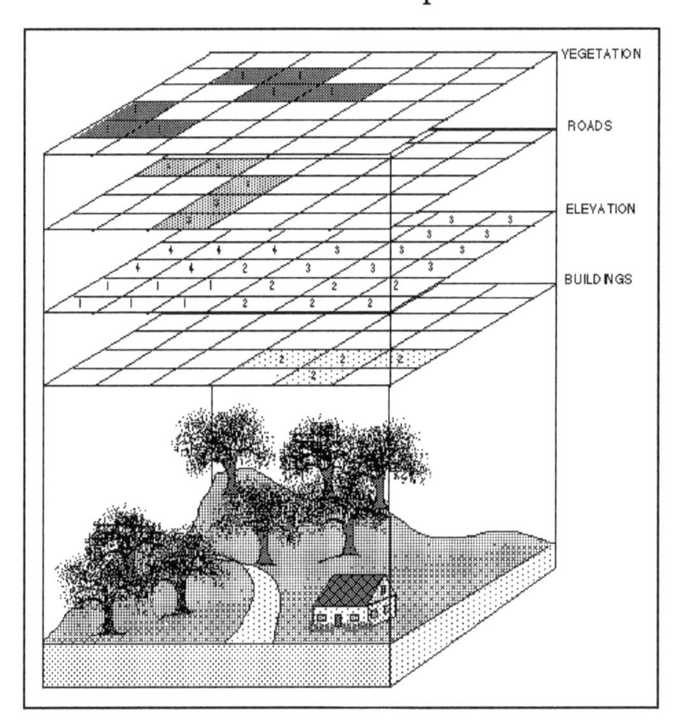

Generally, cells are assigned a single numeric value, but with GRID (a proprietary ArcInfo data format) layers, cell values can also contain additional text and numeric attributes. ArcInfo format grids are the native raster dataset for ArcGIS as well as ArcInfo. In the above diagram, each feature type on the landscape (buildings, elevation, roads, vegetation) is represented in its own raster layer. Note that each raster layer has cells with numbers.

- For the buildings layer, all cell values are 2 (in this case, 2 is a code for houses; other buildings would be encoded with a different value).
- For the elevation layer, the cell value is the elevation at the center of the cell.
- For roads, a value of 3 indicates a road (other road features, e.g., highways, would have a different code).
- For vegetation, trees have a value of 1. In this example, grass is treated as a background value and has no data value (although it could have been given a different numeric value).

All raster datasets are spatially referenced by a very simple method: only one corner of the raster layer is georeferenced. Because cell size is constant in both X and Y directions, cell locations are referenced by row/column designations, rather than with explicit coordinates for the location of each cell's center. This image shows the upper-left corner as the grid origin, with arrows representing the X and Y location of the cells. Different raster file formats may have an origin located at the lower left rather than at the upper left. Each cell or pixel contains a value representing some numerical phenomenon, or a code use for referencing to a non-numerical value.

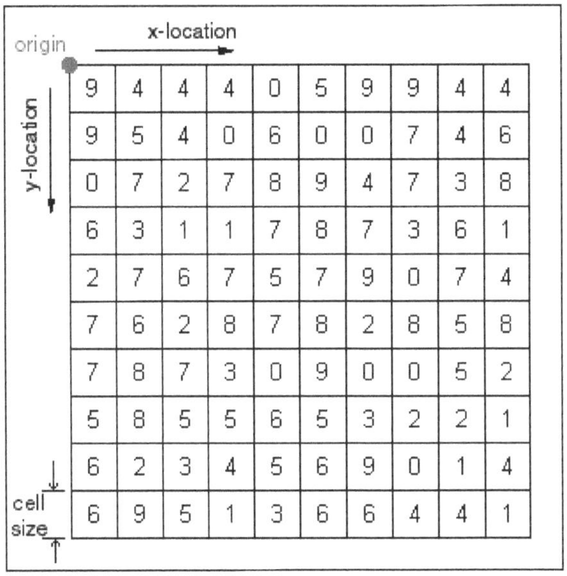

origin
x-location
y-location
cell size

9	4	4	4	0	5	9	9	4	4
9	5	4	0	6	0	0	7	4	6
0	7	2	7	8	9	4	7	3	8
6	3	1	1	7	8	7	3	6	1
2	7	6	7	5	7	9	0	7	4
7	6	2	8	7	8	2	8	5	8
7	8	7	3	0	9	0	0	5	2
5	8	5	5	6	5	3	2	2	1
6	2	3	4	5	6	9	0	1	4
6	9	5	1	3	6	6	4	4	1

Whereas with vector data, each point, node, and vertex has an explicit and absolute coordinate location, raster cells are georeferenced relative to the layer's coordinate origin. This speeds up processing time immensely in comparison to certain types of vector data processing. However, the file sizes of raster datasets can be very large in comparison to vector datasets representing the same phenomenon for the same spatial area. Also, there is a geometric relationship between raster resolution and file size. A raster dataset with cells half as large (e.g., 10 m on a side instead of 20 m on a side) may take up 4 times as much storage space, because it takes four 10 m cells to fit in the space of a single 20 m cell. The following image shows the difference in cell sizes, area, and number of cells for two configurations of the same total area.

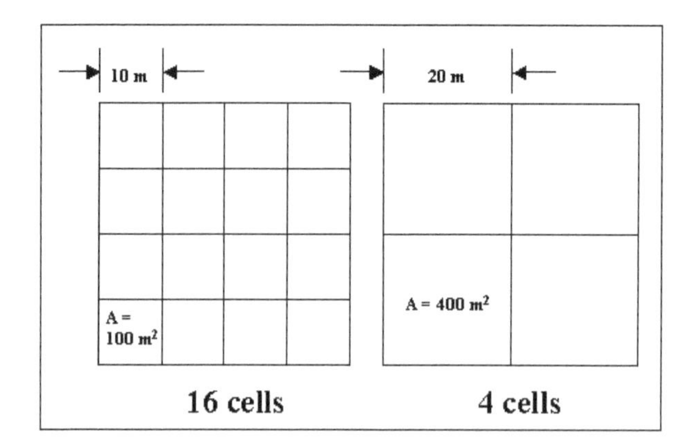

Cells may either have a value (0-infinity) or no value (null, or no data). The difference between these is important. Null values mean that data either fall outside the study area boundary, or that data were either not collected or not available for those cells. In general, when null cells are used in analysis, the output value at a the same cell location is also a null value. Grid datasets can store either integer or floating-point (decimal) data values, though some other data formats can only store integer values. Typical simple image data will have strict limits on the number of unique cell values (typically 0-255).

Pixel or cell? All raster datasets are stored in similar formats. You will want to know the difference between a pixel and a cell, even though they are functionally equivalent. A pixel (short for PICture ELement) represents the smallest resolvable "piece" of a scanned image, whereas a cell represents a user-defined area representing a phenomenon. A pixel is always a cell, but a cell is not always a pixel.

There are many types of raster data you may be familiar with:

- Grids (ArcGIS & ArcInfo specific)
- Graphical images (TIFF, JPEG, BMP, GIF, etc.)
- USGS DEM (Digital Elevation Model)
- Remotely-sensed images (Landsat, SPOT, AVIRIS, AVHRR, Imagine IMG, digital orthophotos)

All raster datasets have essentially the same tessellated structure. Here are few graphical examples of raster data. Note that each image, when zoomed in, shows the same pixellated structure.

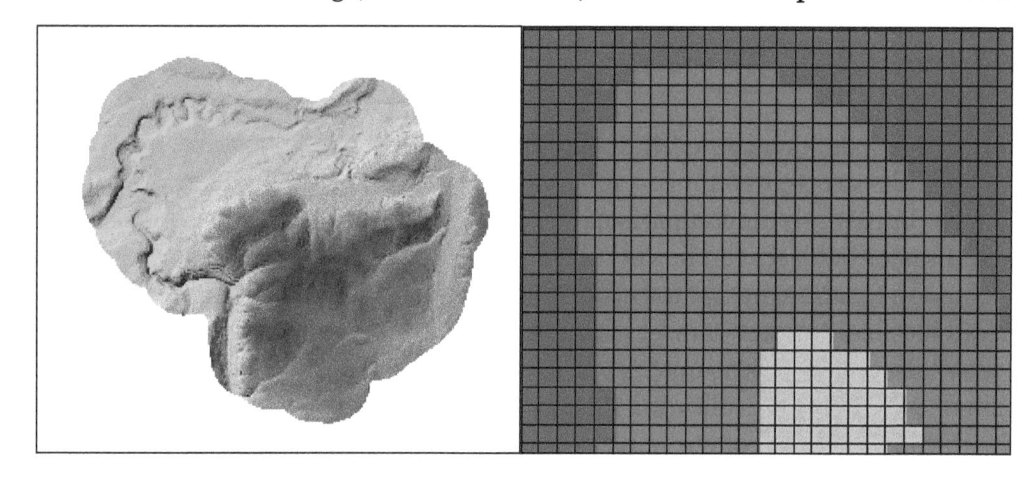

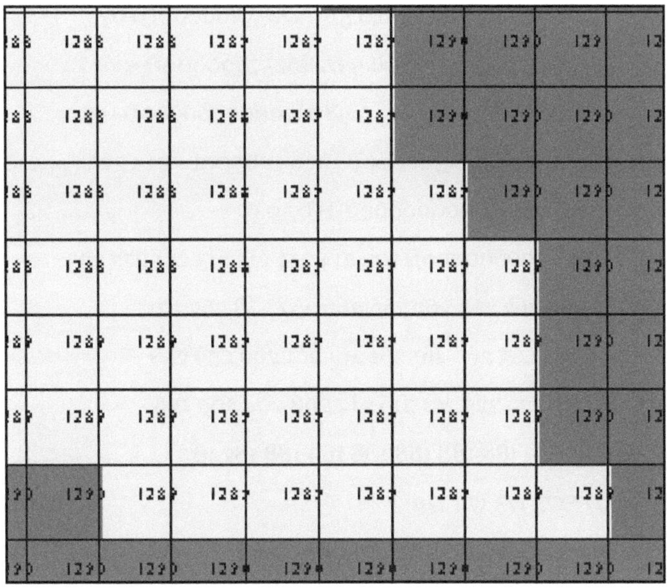

Three differently scaled views of an ArcInfo format elevation grid, showing cell outlines and elevation values.

Most of the raster datasets we will use are single-band, which means that they contain a single "layer" of data. The data can represent elevation, slope, or reflectance.

These single-band images are viewed with a color mapping, so that the cell value is associated with a particular color. For the orthophotos, the color map is a 256-value greyscale ramp. Other raster data, such as elevation models, can be mapped to color ramps that display elevation ranges, as shown in the image directly above. Most GIF files have a limit of 256 unique values (this is known as 8-bit data, because 2^8 = 256).

Multi-band raster data (such as RGB images or satellite images) are generally displayed with a mixture of red, green, and blue values for each different band in the image.

As you can see, when any of the raster layers are displayed at larger scales, the individual cells become visible. As scale of display increases, precision also decreases, and shapes cannot be precisely represented. All spatial datasets are generalized; however, raster datasets more clearly show their level of generalization.

USGS DEMs (Digital Elevation Models) are ASCII (plain text) files which contain georeferencing information as well as point data for elevations on the surface of the earth. Here are the first few lines in the Eatonville, WA 7.5' 30 m DEM:

EATONVILLE, WA 464512215 80000 HAP-81 78-73 08/06/81 BC2 30MX30M INTERVA

EATONVILLE, WA 464512215 FS 1 1 1 10 0.000000000000000D+00

0.000000000000000D+00 0.000000000000000D+00 0.000000000000000D+00

0.000000000000000D+00 0.000000000000000D+00 0.000000000000000D+00

0.000000000000000D+00 0.000000000000000D+00 0.000000000000000D+00

0.000000000000000D+00 0.000000000000000D+00 0.000000000000000D+00

0.000000000000000D+00 0.000000000000000D+00 2 2 4 0.547738572300000D+06

```
0.517735436400000D+07 0.547628092800000D+06 0.519124459100000D+07
0.557153679500000D+06 0.519132801700000D+07 0.557286257000000D+06
0.517743781300000D+07 0.133000000000000D+03 0.987000000000000D+03
0.000000000000000D+00 10.300000E+020.300000E+020.100000E+01 1 322
1 1 92 1 0.547650000000000D+06 0.518850000000000D+07 0.0
0.170000000000000D+03 0.271000000000000D+03 214 214 215 215 221 227 233
240 246 252 257 261 266 264 262 260 262 263 265 266 266 267 268 269 271
270 269 268 268 268 268 267 265 264 263 262 261 261 261 261 260 259 258
258 258 258 256 254 251 245 239 233 229 225 221 216 212 208 204 200 196
194 192 190 190 190 191 190 190 189 189 188 188 188 188 189 188 187 187
187 186 186 184 183 181 180 179 177 175 173 172 170
1 2 218 1 0.547680000000000D+06 0.518472000000000D+07 0.0
0.133000000000000D+03 0.425000000000000D+03 393 396 400 403 408 414 419
421 423 425 423 422 420 412 403 395 391 386 382 373 365 357 347 338 329
324 318 313 312 312 311 308 305 302 301 300 299 297 295 293 289 285 281
279 277 275 270 265 260 253 247 241 233 226 218 216 213 211 209 207 205
205 205 204 202 200 197 197 197 197 197 197 198 199 200 201 206 211 216
218 219 221 216 211 207 186 166 145 142 138 135 135 135 135 135 135 135
134 134 133 134 135 136 135 135 134 148 163 177 185 192 200 202 203 205
206 208 209 210 212 213 213 213 213 213 214 214 214 214 215 219 224 229
235 241 248 253 258 263 262 261 260 261 263 265 266
266 267 268 269 270 269 268 267 267 267 267 265 264 263 262 261 259 260
260 261 259 257 255 255 255 254 252 250 248 243 238 233 228 224 220 216
212 207 204 200 197 195 193 192 191 191 191 190 190 189 189 188 188 188
188 188 188 187 187 186 186 186 184 183 182 180 179 177 176 174 172 171
...
```

The first line lists the file name, data input type (HAP = High Altitude Photography), cell size (30M × 30M). The subsequent lines list elevations for the lattice mesh points (cell centers).

The DEM file is a data source that is not directly usable in most GIS software, but it can be easily imported into ArcGIS, and used for display and analysis.

GEODATA

Geodata is location information stored in a Geographic Information System (GIS). Geodata tackles the problem about location because geographic problems require spatial thinking.

Types of Geographic Data

As it turns out, there's not one single type of geodata. Instead geodata exists in various forms. For example, we commonly use vector and raster to depict geodata.

Vector Files

Vector data consists of vertices and paths. The three basic types of vector data are points, lines and polygons (areas). Each point, line and polygon has a spatial reference frame such as latitude and longitude. First, vector points are simply XY coordinates. Secondly, vector lines connect each point or vertex with paths in a particular order. Finally, polygons join a set of vertices. But it encloses the first and last vertices creating a polygon area.

Raster Files

Raster data is made up of pixels or grid cells. Commonly, they are square and regularly-spaced. But rasters can be rectangular as well. Rasters associate values to each pixel. Continuous rasters have values that gradually change such as elevation or temperature. But discrete rasters set each pixel to a specific class. For example, we represent land cover classes to a set of values.

Geographic Database

The purpose of geographic databases is to house vectors and rasters. Databases store geographic data as a structured set of data/information. For example, Esri geodatabases, geo packages and SpatiaLite are the most common types of geographic databases. We use geographic databases because it's a way to put all data in a single container. Within this container, we can build networks, create mosaics and do versioning.

Web Files

As the internet becomes the largest library in the world, geodata has adapted with its own types of storage and access. For example, GeoJSON, GeoRSS and web mapping services (WMS) were built specifically to serve and display geographic features over the internet. Additionally, online platforms such as Esri's ArcGIS Online allow organizations to build data warehouses in the cloud.

Multi-temporal

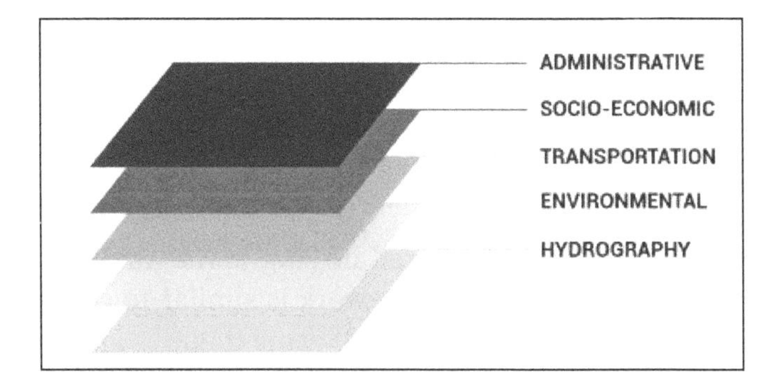

Multi-temporal data attaches a time component to information. But multi-temporal geodata not only has a time component, but a geographic component as well. For example, weather and climate data tracks how temperature and meteorological information changes in time in a geographical context. Other examples of multi-temporal geodata are demographic trends, land use patterns and lightning strikes.

The truth is:

You can group geodata into as many themes as you want. They can be as broad or as narrow to your liking.

Here are examples of geographic themes:

Cultural

- Administrative (Boundaries, cities and planning).
- Socioeconomic data (Demographics, economy and crime).
- Transportation (Roads, railways and airport).

Physical

- Environmental data (Agriculture, soils and climate).
- Hydrography data (Oceans, lakes and rivers).
- Elevation data (Terrain and relief).

Sources for Geodata

Are you trying to find open, authoritative geodata to use in your maps?

Before the concept of open data took off, organizations were protecting data as if it was Fort Knox. Since then, we are in a much better position.

Currently, there's no single website that holds all the geodata in the world. Instead they branch out into what they are most specialized in.

For example, Open Street Map data is the largest crowd-sourced GIS database in the world providing countless applications for the public.

GEOMETRIC NETWORKS

A geometric network is an object commonly used in geographic information systems to model a series of interconnected features. A geometric network is similar to a graph in mathematics and computer science, and can be described and analyzed using theories and concepts similar to graph theory. Geometric networks are often used to model road networks and public utility networks (such as electric, gas, and water utilities).

Composition of a Geometric Network

A geometric network is composed of edges that are connected. Connectivity rules for the network specify which edges are connected and at what points they are connected, commonly referred to as junction or intersection points. These edges can have weights or flow direction assigned to them, which dictate certain properties of these edges that affect analysis results. In the case of certain types of networks, source points (points where flow originates) and sink points (points where flow terminates) may also exist. In the case of utility networks, a source point may correlate with an electric substation or a water pumping station, and a sink point may correlate with a service connection at a residential household.

Functions

Networks define the interconnectedness of features. Through analyzing this connectivity, paths from one point to another on the network can be traced and calculated. Through optimization algorithms and utilizing network weights and flow, these paths can also be optimized to show specialized paths, such as the shortest path between two points on the network, as is commonly done in the calculation of driving directions. Networks can also be used to perform spatial analysis to determine points or edges that are encompassed in a certain area or within a certain distance of a specified point. This has applications in hydrology and urban planning, among other fields.

Applications

- Routing: For calculating driving directions, paths from one point of interest to another, locating nearby points of interest.
- Urban Planning: For site suitability studies, and traffic and congestion studies.
- Electric Utility Industry: For modeling an electrical grid in GIS, tracing from a generation source.
- Other Public Utilities: For modeling water distribution flow and natural gas distribution.

GIS MAPPING

GIS mapping helps you to visualize and identify patterns that are difficult to see if the data elements are in table format. It also helps to identify patterns that emerge when you view two or more datasets together.

What can be Mapped?

You can use GIS to map any data element that can be tied to a latitude and longitude (a geospatial point). In business, this may include the locations of current customers, the locations of consumers who have a high likelihood of becoming customers, the locations of competitors, estimates of demand for specific services per household, demographic characteristics such as average income, and more.

How is GIS Mapping Used?

Since GIS mapping technology allows you to turn data layers on and off, it can be used either to focus on specific data elements or to view new combinations of elements. Regardless of the approach, the goal is to identify patterns that can inform business decisions.

For example, a healthcare system strategist may use GIS mapping to view estimated demand for specific service lines layered over their organization's network of locations. By looking at areas of high and low demand in relation to existing facilities, they may spot opportunities to optimize which service lines are offered at each location.

A retail or restaurant site selector may use GIS technology to view potential customer density, competitor locations, and other area draw factors to determine the viability of a potential site.

In many cases, the geographic information system allows users to run reports on the underlying data and can be used as the base for other analytics tools, such as site scoring tools.

GIS Mapping for Environmental Management

GIS, has a very important role to play in environmental management. GIS, GPS and remote sensing technologies provide tools to researchers to help them to understand, visualize, intergrate and quantify available research data.

Land cover maps and databases can be created from satellite imagery or aerial photography and are then stored and digitized (feature extraction) to create geospatial databases in a GIS environment.

Feature extraction services include structures and facilities (buildings, bridges, airports, etc.), land use, land cover, vegetation classification (agriculture, forestry, environmental, mineral), water bodies, well locations, soil types, pipelines and more.

GIS assists environmental managers the production of GIS maps that can show how our natural resources respond to change over time including coastal, vegetation and geological.

GIS Map integrated with Pleiades 3D ArcScene (0.5m). Ditigal Terrain Model (DTM) 1 meter Bingham Canyon Copper Mine, Utah, USA.

GIS Mapping for Wildlife Conservation

GIS, GPS and remote sensing technology provides wildlife conservationists and researchers with informed management decisions to help identify habitats and protect species to assist in decision making where conservation efforts are needed the most.

High resolution satellite imagery can give researchers increasingly up-to-date geospatial data and information for monitoring wildlife migrations, poaching activities, habitat mapping and tracking endangered species in remote areas of the world to assist in management and conservation activities.

Wildlife Monitoring

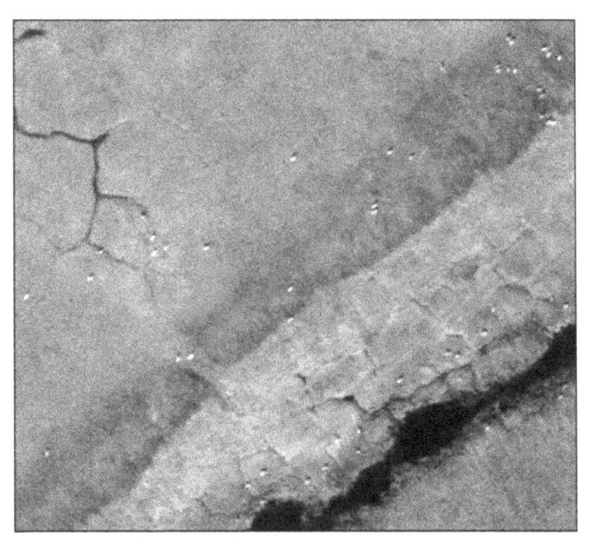

WorldView-3 Satellite Image (30cm) Monitoring Alaskan Caribou.

GIS technology allows researchers to track wildlife in the remote areas of the world. The GIS data collected from GPS and high resolution satellite imagery or unmanned aerial systems allows researchers to identify and monitor wildlife movement, patterns, species numbers, behaviors and to assist in the prevention of poaching.

Successful applications in GIS technology for wildlife management has demonstrated how this tool can provide valuable information to researchers and conservationists.

Web Mapping

Web mapping is the process of using the maps delivered by geographic information systems (GIS) in World Wide Web. A web map on the World Wide Web is both served and consumed, thus web mapping is more than just web cartography, it is a service by which consumers may choose what the map will show. Web GIS emphasizes geodata processing aspects more involved with design aspects such as data acquisition and server software architecture such as data storage and algorithms, than it does the end-user reports themselves.

The terms web GIS and web mapping remain somewhat synonymous. Web GIS uses web maps, and end users who are web mapping are gaining analytical capabilities. The term location-based services refers to web mapping consumer goods and services. Web mapping usually involves a web browser or other user agent capable of client-server interactions. Questions of quality, usability, social benefits, and legal constraints are driving its evolution.

The advent of web mapping can be regarded as a major new trend in cartography. Until recently cartography was restricted to a few companies, institutes and mapping agencies, requiring relatively expensive and complex hardware and software as well as skilled cartographers and geomatics engineers.

Web mapping has brought many geographical datasets, including free ones generated by OpenStreetMap and proprietary datasets owned by HERE, Google, Tencent, TomTom, and others. A range of free software to generate maps has also been conceived and implemented alongside proprietary tools like ArcGIS. As a result, the barrier to entry for serving maps on the web has been lowered.

Types of Web Maps

A first classification of web maps has been made by Kraak in 2001. He distinguished *static* and *dynamic* web maps and further distinguished *interactive* and *view only* web maps. Today there an increased number of dynamic web maps types, and static web map sources.

Analytical Web Maps

Analytical web maps offer GIS analysis. The geodata can be a static provision, or needs updates. The borderline between analytical web maps and web GIS is fuzzy. Parts of the analysis can be carried out by the GIS geodata server. As web clients gain capabilities processing is distributed.

Animated and Realtime

Realtime maps show the situation of a phenomenon in close to realtime (only a few seconds or minutes delay). They are usually animated. Data is collected by sensors and the maps are generated or updated at regular intervals or on demand.

Animated maps show changes in the map over time by animating one of the graphical or temporal variables. Technologies enabling client-side display of animated web maps include scalable vector graphics (SVG), Adobe Flash, Java, QuickTime, and others. Web maps with real-time animation include weather maps, traffic congestion maps and vehicle monitoring systems.

CartoDB launched an open source library, Torque, which enables the creation of dynamic animated maps with millions of records. Twitter uses this technology to create maps to reflect how users reacted to news and events worldwide.

Collaborative Web Maps

Collaborative maps are a developing potential. In proprietary or open source collaborative software, users collaborate to create and improve the web mapping experience. Some collaborative web mapping projects are:

- Google Map Maker,
- Here Map Creator,
- OpenStreetMap,

- WikiMapia,
- Meta: Maps - a survey of Wikimedia movement web mapping proposals.

Online Atlases

The traditional atlas goes through a remarkably large transition when hosted on the web. Atlases can cease their printed editions or offer printing on demand. Some atlases also offer raw data downloads of the underlying geospatial data sources.

Static Web Maps

Static web pages are *view only* without animation or interactivity. These files are created once, often manually, and infrequently updated. Typical graphics formats for static web maps are PNG, JPEG, GIF, or TIFF (e.g., drg) for raster files, SVG, PDF or SWF for vector files. These include scanned paper maps not designed as screen maps. Paper maps have a much higher resolution and information density than typical computer displays of the same physical size, and might be unreadable when displayed on screens at the wrong resolution.

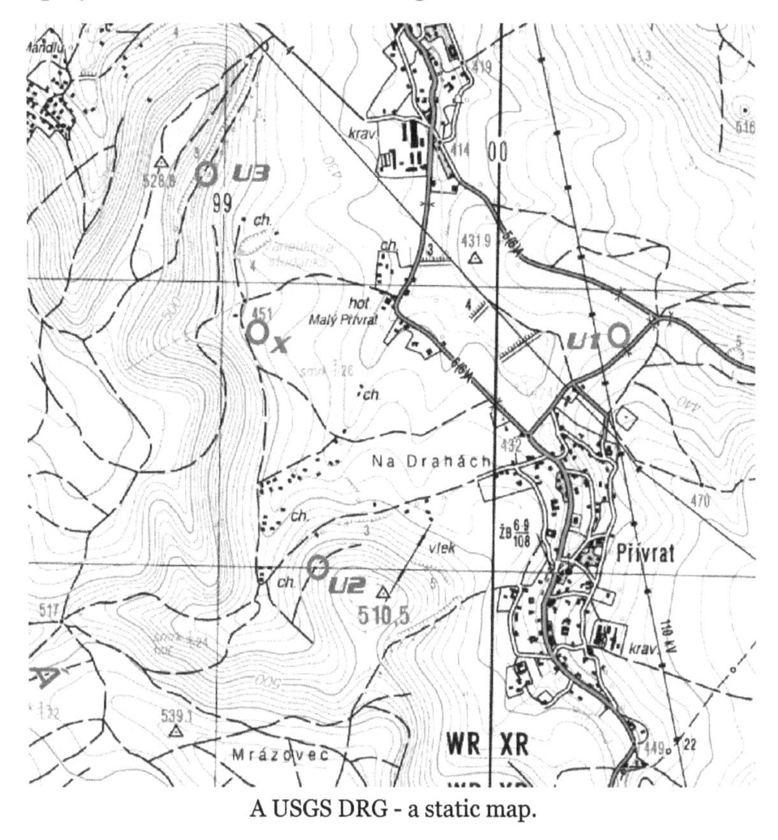

A USGS DRG - a static map.

Web GIS in the Cloud

Various companies now offer web mapping as a cloud based software as a service. These service providers allow users to create and share maps by uploading data to their servers (cloud storage). The maps are created either by using an in browser editor or writing scripts that leverage the service providers API's.

Evolving paper Cartography

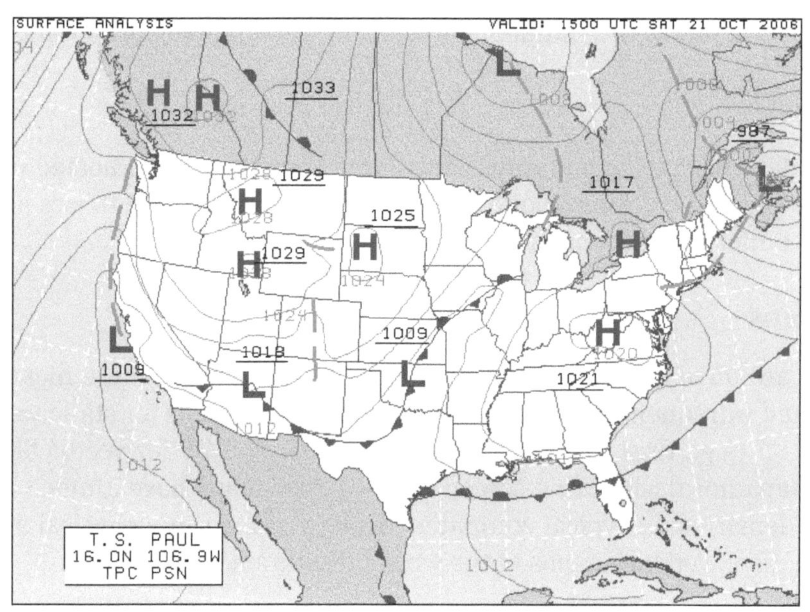

A surface weather analysis for the United States on October 21, 2006.

Compared to traditional techniques, mapping software has many advantages. The disadvantages are also stated.

- Web maps can easily deliver up to date information: If maps are generated automatically from databases, they can display information in almost realtime. They don't need to be printed, mastered and distributed. Examples:
    - A map displaying election results, as soon as the election results become available.
    - A traffic congestion map using traffic data collected by sensor networks.
    - A map showing the current locations of mass transit vehicles such as buses or trains, allowing patrons to minimize their waiting time at stops or stations, or be aware of delays in service.
    - Weather maps, such as NEXRAD.
- Software and hardware infrastructure for web maps is cheap: Web server hardware is cheaply available and many open source tools exist for producing web maps. Geodata, on the other hand, is not; satellites and fleets of automobiles use expensive equipment to collect the information on an ongoing basis. Perhaps owing to this, many people are still reluctant to publish geodata, especially in places where geodata are expensive. They fear copyright infringements by other people using their data without proper requests for permission.
- Product updates can easily be distributed: Because web maps distribute both logic and data with each request or loading, product updates can happen every time the web user reloads the application. In traditional cartography, when dealing with printed maps or interactive maps distributed on offline media (CD, DVD, etc.), a map update takes serious efforts, triggering a reprint or remastering as well as a redistribution of the media. With web maps,

data and product updates are easier, cheaper, and faster, and occur more often. Perhaps owing to this, many web maps are of poor quality, both in symbolization, content and data accuracy.

- Web maps can combine distributed data sources: Using open standards and documented APIs one can integrate (*mash up*) different data sources, if the projection system, map scale and data quality match. The use of centralized data sources removes the burden for individual organizations to maintain copies of the same data sets. The downside is that one has to rely on and trust the external data sources. In addition, with detailed information available and the combination of distributed data sources, it is possible to find out and combine a lot of private and personal information of individual persons. Properties and estates of individuals are now accessible through high resolution aerial and satellite images throughout the world to anyone.
- Web maps allow for personalization: By using user profiles, personal filters and personal styling and symbolization, users can configure and design their own maps, if the web mapping systems supports personalization. Accessibility issues can be treated in the same way. If users can store their favourite colors and patterns they can avoid color combinations they can't easily distinguish (e.g. due to color blindness). Despite this, as with paper, web maps have the problem of limited screen space, but more so. This is in particular a problem for mobile web maps; the equipment carried usually has a very small screen, making it less likely that there is room for personalisation.
- Web maps enable collaborative mapping: Similar to web mapping technologies such as DHTML/Ajax, SVG, Java, Adobe Flash, etc. enable distributed data acquisition and collaborative efforts. Examples for such projects are the OpenStreetMap project or the Google Earth community. As with other open projects, quality assurance is very important, however, and the reliability of the internet and web server infrastructure is not yet good enough. Especially if a web map relies on external, distributed data sources, the original author often cannot guarantee the availability of the information.
- *Web maps support hyperlinking to other information on the web*: Just like any other web page or a wiki, web maps can act like an index to other information on the web. Any sensitive area in a map, a label text, etc. can provide hyperlinks to additional information. As an example a map showing public transport options can directly link to the corresponding section in the online train time table. However, development of web maps is complicated enough as it is: Despite the increasing availability of free and commercial tools to create web mapping and web GIS applications, it is still a more complex task to create interactive web maps than to typeset and print images. Many technologies, modules, services and data sources have to be mastered and integrated The development and debugging environments of a conglomerate of different web technologies is still awkward and uncomfortable.

Web Mapping Technologies

Web mapping technologies require both server-side and client-side applications. The following is a list of technologies utilized in web mapping:

- Spatial databases are usually object relational databases enhanced with geographic data types, methods and properties. They are necessary whenever a web mapping application has to deal with dynamic data (that changes frequently) or with huge amount of geographic data. Spatial databases allow spatial queries, sub selects, reprojections, and geometry manipulations and offer various import and export formats. PostGIS is a prominent example; it is open source. MySQL also implements some spatial features. Oracle Spatial, Microsoft SQL Server (with the spatial extensions), and IBM DB2 are the commercial alternatives. The Open Geospacial Consortium's (OGC) specification "Simple Features" is a standard geometry data model and operator set for spatial databases. Part 2 of the specification defines an implementation using SQL.

- Tiled web maps display rendered maps made up of raster image "tiles".
- Vector tiles are also becoming more popular—Google and Apple have both transitioned to vector tiles. Mapbox.com also offers vector tiles. This new style of web mapping is resolution independent, and also has the advantage of dynamically showing and hiding features depending on the interaction.
- WMS servers generate maps using parameters for user options such as the order of the layers, the styling and symbolization, the extent of the data, the data format, the projection, etc. The OGC standardized these options. Another WMS server standard is the Tile Map Service. Standard image formats include PNG, JPEG, GIF and SVG. Open source WMS Servers include UMN Mapserver, GeoServer and Mapnik. Commercial alternatives exist from most commercial GIS vendors, such as ESRI ArcIMS and CadCorp.

Collaborative Mapping

Collaborative mapping is the aggregation of Web mapping and user-generated content, from a group of individuals or entities, and can take several distinct forms. With the growth of technology for storing and sharing maps, collaborative maps have become competitors to commercial services, in the case of OpenStreetMap, or components of them, as in Google Map Maker and Yandex. Map editor.

Volunteers collect geographic information and the citizens/individuals can be regarded as sensors within a geographical environment that create, assemble, and disseminate geographic data provided voluntarily by the individuals. Collaborative mapping is a special case of the larger phenomenon known as crowd sourcing, that allows citizens to be part of collaborative approach to accomplish a goal. The goals in collaborative mapping have a geographical aspect, e.g. having a more active role in urban planning. Especially when data, information, knowledge is distributed in a population and an aggregation of data is not available, then collaborative mapping can bring a benefit for the citizens or activities in a community with an e-Planing Platform. Extensions of critical and participatory approaches to geographic information systems combines software tools with a joint activities to accomplish a community goal. Additionally, the aggregated data can be used for a Location-based service like available public transport options at the geolocation where a mobile device is currently used (GPS-sensor). The relevance for the user at a specific geolocation cannot be represented with logic value in general (relevant=true/false). The relevance can be represented with Fuzzy-Logic or a Fuzzy architectural spatial analysis.

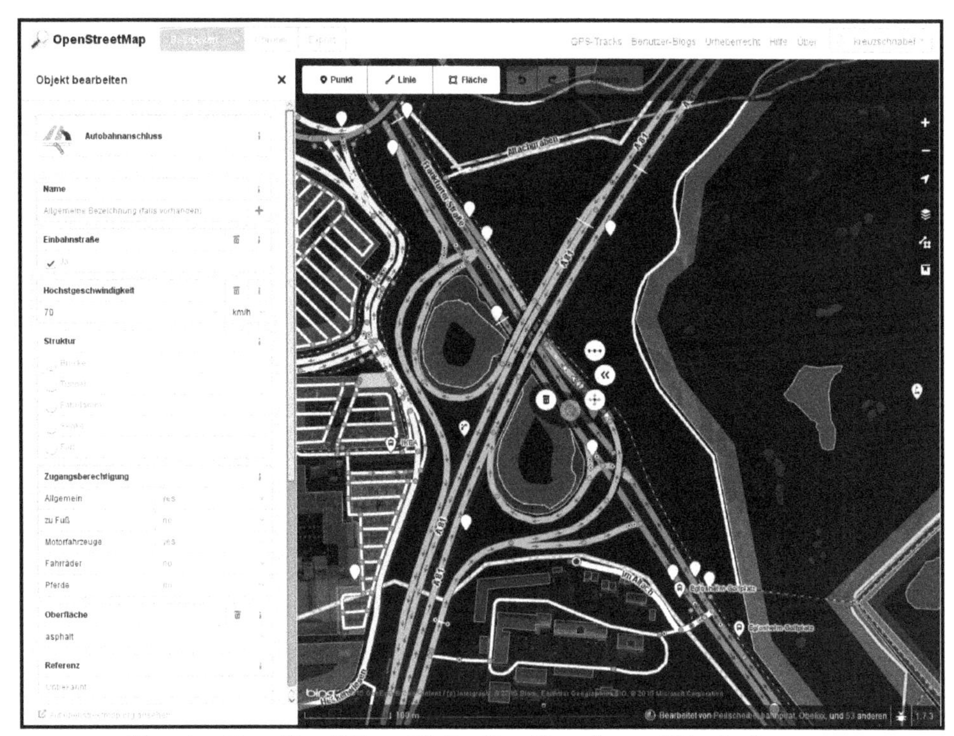

Mapping OpenStreetMap with the iD editor, done within the browser.

Types

Collaborative mapping applications vary depending on which feature the collaborative edition takes place: on the map itself (shared surface), or on overlays to the map. A very simple collaborative mapping application would just plot users' locations (social mapping or geosocial networking) or Wikipedia articles' locations (Placeopedia). Collaborative implies the possibility of edition by several distinct individuals so the term would tend to exclude applications where the maps are not meant for the general user to modify.

In this kind of application, the map itself is created collaboratively by sharing a common surface. For example, both OpenStreetMap and WikiMapia allow for the creation of single 'points of interest', as well as linear features and areas. Collaborative mapping and specifically surface sharing faces the same problems as revision control, namely concurrent access issues and versioning. In addition to these problems, collaborative maps must deal with the difficult issue of cluttering, due to the geometric constraints inherent in the media. One approach to this problem is using overlays, allowing to suitable use in consumer services. Despite these issues, collaborative mapping platforms such as OpenStreetMap can be considered as being as trustworthy as professionally produced maps.

Overlays group together items on a map, allowing the user of the map to toggle the overlay's visibility and thus all items contained in the overlay. The application uses map tiles from a third-party (for example one of the mapping APIs) and adds its own collaboratively edited overlays to them, sometimes in a wiki fashion. If each user's revisions are contained in an overlay, the issue of revision control and cluttering can be mitigated. One example of this is the accessibility platform Accessadvisr, which utilises collaborative mapping to inform persons of accessibility issues, which is perceived to be as reliable and trustworthy as professional information.

Other overlays-based collaborative mapping tools follow a different approach and focus on user centered content creation and experience. There users enrich maps with their own points of interest and build kind of travel books for themselves. At the same time users can explore overlays of other users as collaborative extension.

Humanitarian Collaborative Mapping

Humanitarian OpenStreetMap Team, based on OpenStreetMap, provides collaborative mapping support for humanitarian objectives, e.g. collaborative transportation map, epidemiological mapping for Malaria, earthquake response, or typhoon response.

Private Local Collaboration using Maps

Some mapping companies offer an online mapping tool that allows private collaboration between users when mapping sensitive data on digital maps, e.g.:

- Google Maps.
- Wegovnow: a map based platform to engage the local civic society – local collaboration & publishing with maps.

Quality Assurance

If citizens or a community collects data, information then concerns come up about data quality, and specifically about its credibility. The same aspects of quality assurance are relevant for collaborative mapping and the possibility of vandalism.

Data Collection Tools

Collaborative mapping is not restricted to the application of mobile devices but if data is captured with a mobile device the satellite navigation (like GPS is helpful to assign the current geolocation to the collected data at the geolocation. Open Source tools like Open Data Kit are used to collect the mapping data (e.g. about health care facilities or humanitarian operations) with a survey that could automatically insert the geolocation into the survey data that could include visual information (e.g. images, videos) and audio samples collected at the current geolocation. An image can be used e.g. as additional information of damage assessment after an earth quake.

Restricted Visibility of Alterations

These sites provide general base map information and allow users to create their own content by marking locations where various events occurred or certain features exist, but aren't already shown on the base map. Some examples include 311-style request systems and 3D spatial technology.

Public Alterations and Quality Assured Versions

The openness for changes to the community is possible for all individuals and the community is validating changes by putting regions and location at their personal watchlist. Any changes in the joint repository of the mapping process are captured by a version control system- Reverting changes is possible and specific quality assured versions of specific areas can be marked as reference map for a specific area. Quality assurance can be implemented on different scales:

- Version of complete map,
- Version of selected regions/area,
- Version of mapping attributes a Point of Interest (e.g. hospital marked as "under construction" is providing health care services).

Blockchain can be used as integrity check of alterations or digital signature can be used to mark a certain version as "quality assured" by the institution that signed a map as digital file or digital content.

Heat Maps in GIS

Heat mapping, from a geographic perspective, is a method of showing the geographic clustering of a phenomenon. Sometimes also referred as hot spot mapping, heat maps show locations of higher densities of geographic entities. The 'heat' in the term refers to the concentration of the geographic entity within any given spot, not to be confused with heat mapping that refers to the mapping of actual temperatures on the earth's surface. Heat mapping is a way of geographically visualizing locations so that patterns of higher than average occurrence of things likes crime activity, traffic accidents, or store locations can emerge.

One way to create a heat map is by interpolating discrete points to create a continuous surface known as a density surface. When calculating a density surface, three main parameters have to be determined that will affect the results. Given that the output is a raster file, the cell size will be a determining factor as to the degree of detail in terms of coarseness of the resulting density surface. The larger the cell size, the more of a staircase effect on the resulting surface layer. Conversely, a smaller cell size will result in a smoother surface but processing will take longer and will result in a larger file size. The suggested balance is to set the cell size between 10 and 100 cells per density unit.

The bandwidth or search radius is the second parameter to be set. The search radius is the area around each cell the GIS software will factor into the density calculation. Set a search radius too small and density patterns will be restricted to the immediate area of the point features. Set a search radius too large and density patterns will become too generalized.

The third parameter is the type of calculation used in interpolating the density surface. The most simple calculation is a straightforward count of features within a search radius. The more common calculation is to use a weighted calculation such as Inverse distance weighting (IDW). IDW assigns more weight to features closest to the starting cell than to features farther away. In other words, the weight of a given point is in inverse proportion to its distance from the interpolated cell.

The resulting density surface is visualized using a gradient that allows the areas of highest density (or hot spots) to be easily identified. This heat map was created from geotagged photos from Panoramio by Ahti Heinla who created what he named the "World touristiness map." Yellow indicates areas with the highest concentrations of photos taken by the highest number of individual photographers. Red indicates areas of moderate "touristiness" and blue are the lowest levels. Grey areas have no photos in Panoramio. The areas of highest density in terms of photos uploaded are immediately identifiable with Europe, areas along the coasts in the United States, and areas within Asia most notable.

Heat maps are particularly popular for crime prevention planning by law enforcement agencies as being able to identify discernible geographic clusters of higher criminal activity allows for more intelligent deployment of police resources to areas of high crime. The creation of heat maps has other applications besides crime mapping. In the example above, a heat map was created to show worldwide "touristiness" based on geotagged images uploaded to Flickr.

Heat Mapping Tools

Heatmap.py is a python script for generating heat maps based on coordinate data.

Gheat "implements a map tile server for a heatmap layer."

As the name implies, HeatMapAPI is an API (with both a limited free and licensed version) that integrates heat map images into Google Maps.

Google Fusion Tables has a heat map function available as part of the options for visualizing geographic data.

Heatmap.js generates web heatmaps with the html5 canvas element created by Patrick Wied.

Digital Geological Mapping

Digital geologic mapping is the process by which geological features are observed, analyzed, and recorded in the field and displayed in real-time on a computer or personal digital assistant (PDA). The primary function of this emerging technology is to produce spatially referenced geologic maps that can be utilized and updated while conducting field work.

Screenshot of a structure map generated by geological mapping software for an 8500ft deep gas & Oil reservoir in the Erath field, Vermilion Parish, Erath, Louisiana. The left-to-right gap, near the top of the contour map indicates a Fault line. This fault line is between the blue/green contour lines and the purple/red/yellow contour lines. The thin red circular contour line in the middle of the map indicates the top of the oil reservoir. Because gas floats above oil, the thin red contour line marks the gas/oil contact zone.

Traditional Geologic Mapping

Geologic mapping is an interpretive process involving multiple types of information, from analytical data to personal observation, all synthesized and recorded by the geologist. Geologic observations have traditionally been recorded on paper, whether on standardized note cards, in a notebook, or on a map.

Mapping in the Digital Era

In the 21st century, computer technology and software are becoming portable and powerful enough to take on some of the more mundane tasks a geologist must perform in the field, such as precisely locating oneself with a GPS unit, displaying multiple images (maps, satellite images, aerial photography, etc.), plotting strike and dip symbols, and color-coding different physical characteristics of a lithology or contact type (e.g., unconformity) between rock strata. Additionally, computers can now perform some tasks that were difficult to accomplish in the field, for example, handwriting or voice recognition and annotating photographs on the spot.

Digital mapping has positive and negative effects on the mapping process; only an assessment of its impact on a geological mapping project as a whole shows whether it provides a net benefit. With the use of computers in the field, the recording of observations and basic data management changes dramatically. The use of digital mapping also affects when data analysis occurs in the mapping process, but does not greatly affect the process itself.

Advantages

- Data entered by a geologist may have fewer errors than data transcribed by a data entry clerk.
- Data entry by geologists in the field may take less total time than subsequent data entry in the office, potentially reducing the overall time needed to complete a project.
- The spatial extent of real world objects and their attributes can be entered directly into a database with geographic information system (GIS) capability. Features can be automatically color-coded and symbolized based on set criteria.
- Multiple maps and imagery (geophysical maps, satellite images, orthophotos, etc.) can easily be carried and displayed on-screen.
- Geologists may upload each other's data files for the next day's field work as reference.
- Data analysis may start immediately after returning from the field, since the database has already been populated.
- Data can be constrained by dictionaries and dropdown menus to ensure that data are recorded systematically and that mandatory data are not forgotten.
- Labour-saving tools and functionality can be provided in the field e.g. structure contours on the fly, and 3D visualisation.
- Systems can be wirelessly connected to other digital field equipment (such as digital cameras and sensor webs).

Disadvantages

- Computers and related items (extra batteries, stylus, cameras, etc.) must be carried in the field.
- Field data entry into the computer may take longer than physically writing on paper, possibly resulting in longer field programs.
- Data entered by multiple geologists may contain more inconsistencies than data entered by one person, making the database more difficult to query.
- Written descriptions convey to the reader detailed information through imagery that may not be communicated by the same data in parsed format.

- Geologists may be inclined to shorten text descriptions because they are difficult to enter (either by handwriting or voice recognition), resulting in loss of data.
- There are no original, hardcopy field maps or notes to archive. Paper is a more stable medium than digital format.

Educational and Scientific Uses

Some universities and secondary educators are integrating digital geologic mapping into class work. For example, The GeoPad project describes the combination of technology, teaching field geology, and geologic mapping in programs such as Bowling Green State University's geology field camp. At Urbino University (Italy) it: Università di Urbino, Field Digital Mapping Techniques are integrated in Earth and Environmental Sciences courses since 2006. The MapTeach program is designed to provide hands-on digital mapping for middle and high school students. The SPLINT project in the UK is using the BGS field mapping system as part of their teaching curriculum

Digital mapping technology can be applied to traditional geologic mapping, reconnaissance mapping, and surveying of geologic features. At international digital field data capture (DFDC) meetings, major geological surveys (e.g., British Geological Survey and Geological Survey of Canada) discuss how to harness and develop the technology. Many other geological surveys and private companies are also designing systems to conduct scientific and applied geological mapping of, for example, geothermal springs and mine sites.

Equipment

The initial cost of digital geologic computing and supporting equipment may be significant. In addition, equipment and software must be replaced occasionally due to damage, loss, and obsolescence. Products moving through the market are quickly discontinued as technology and consumer interests evolve. A product that works well for digital mapping may not be available for purchase the following year; however, testing multiple brands and generations of equipment and software is prohibitively expensive.

Common Essential Features

Some features of digital mapping equipment are common to both survey or reconnaissance mapping and "traditional" comprehensive mapping. The capture of less data-intensive reconnaissance mapping or survey data in the field can be accomplished by less robust databases and GIS programs, and hardware with a smaller screen size.

- Devices and software are intuitive to learn and easy to use.
- Rugged, as typically defined by military standards (MIL-STD-810) and ingress protection ratings.
- Waterproof.
- Screen is easy to read in bright sunlight and on gray sky days.
- Removable static memory cards can be used to back up data.

- Memory on board is recoverable.
- Real-time and post-processing differential correction for GPS locations.
- Portable battery with at least 9 hours of life at near constant use.
- Can change batteries in the field.
- Batteries should have no "memory," such as with NiCd.
- Chargeable by unconventional power sources (generators, solar, etc.)
- Wireless real-time link to GPS or built-in GPS.
- Wireless real-time link from computer to camera and other peripherals.
- USB port(s).

Features Essential to Capture Traditional Geologic Observations

Hardware and software only recently (in 2000) became available that can satisfy most of the criteria necessary for digitally capturing "traditional" mapping data.

- Screen about 5" x 7"—compact but large enough to see map features. In 2009, some traditional mapping is conducted on PDAs.
- Lightweight—ideally less than 3 lbs.
- Transcription to digital text from handwriting and voice recognition.
- Can store paragraphs of data (text fields).
- Can store complex relational database with drop-down lists.
- Operating system and hardware are compatible with a robust GIS program.
- At least 512 MB memory.

Software

Since every geologic mapping project covers an area with unique lithologies and complexities, and every geologist has a unique style of mapping, no software is perfect for digital geologic mapping out of the box. The geologist can choose to either modify their mapping style to the available software, or modify the software to their mapping style, which may require extensive programming. As of 2009, available geologic mapping software requires some degree of customization for a given geologic mapping project. Some digital-mapping geologists/programmers have chosen to highly customize or extend ESRI's ArcGIS instead. At digital field data capture meetings such as at the British Geological Survey in 2002 some organisations agreed to share development experiences, and some software systems are now available to download for free.

Production Mapping

Production mapping focuses on standardizing and streamlining the four main workflows of GIS production: data capture, editing, review, and cartographic output. Mapping, charting, and visualization tools focus on providing standardized methods for users to create and maintain cartographic outputs from a GIS.

Templates

ESRI Production Mapping cartography tools allow you to define templates for mapping tasks. Templates are reused in different geographic areas or used to maintain and update existing maps or charts.

Workflows

Mapping and charting tools support enterprise cartographic production workflows. These workflows generate numerous cartographic products. Groups of users maintain the cartographic products. Typically, maps and charts in these products have a high degree of detail and a standardized look and feel. Products are maintained over time and require updates to their content as their geographic areas change.

Cartographic output workflows can be subdivided into smaller workflows and tasks. ESRI Production Mapping has tools to support the following workflows.

Context

What is the purpose of the map? What geographic area will it cover? What scale, projection, and data will be used? ESRI Production Mapping helps automate and store these settings to ensure consistency and integrity of the products.

Cartographic Data

Database-driven cartography allows the automation of maps and charts based on the information stored with the GIS. A suite of tools extend traditional GIS data with features and required attributes and enhance cartographic workflow.

Symbology

Symbology workflows link feature attributes (data dictionary) to symbols and maintain that rule base inside a geodatabase. Shared map templates and groups of layer files ensure that your team produces standardized products.

Cartographic Editing

Interactive and automated tools support cartographic editing of GIS data and map production workflows.

Layout

ESRI Production Mapping provides an enhanced environment where the page layout can be managed and manipulated more effectively. Additional surround elements extend the type of information that can be displayed in a page layout.

Publishing

Maps and charts can be printed, exported, and archived in standard formats. This includes the ability to support color separation options.

GIS FORMATS AND GEOSPATIAL FILE EXTENSIONS

Vector GIS File Formats

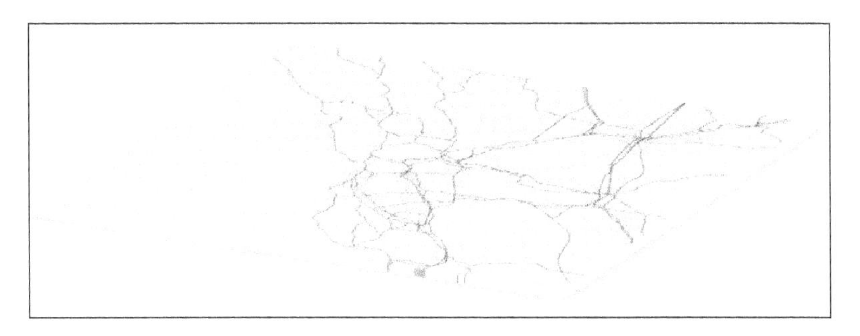

Vector data is not made up of grids of pixels. Instead, vector graphics are comprised of vertices and paths.

The three basic symbol types for vector data are points, lines and polygons (areas). These GIS file formats house vector data.

Esri Shape File

SHP,.DBF,.SHX:

The shape file is BY FAR the most common geospatial file type you'll encounter. All commercial and open source accept shape file as a GIS format. It's so ubiquitous that it's become the industry standard.

But you'll need a complete set of three files that are mandatory to make up a shape file. The three required files are:

- SHP is the feature geometry.
- SHX is the shape index position.
- DBF is the attribute data.

You can optionally include these files but are not completely necessary.

- PRJ is the projection system metadata.
- XML is the associated metadata.
- SBN is the spatial index for optimizing queries.
- SBX optimizes loading times.

Geographic JavaScript Object Notation (GeoJSON)

GEOJSON.JSON:

The GeoJSON format is mostly for web-based mapping. GeoJSON stores coordinates as text in JavaScript Object Notation (JSON) form. This includes vector points, lines and polygons as well as tabular information.

GeoJSON store objects within curly braces and in general have less markup overhead (compared to GML). GeoJSON has straightforward syntax that you can modify in any text editor.

Webmaps browsers understand JavaScript so by default GeoJSON is a common web format. But JavaScript only understands binary objects. Fortunately, JavaScript can convert JSON to binary.

Geography Markup Language (GML)

.GML:

GML allows for the use of geographic coordinates extension of XML. And eXtensible Markup Language (XML) is both human-readable and machine-readable.

GML stores geographic entities (features) in the form of text. Similar to GeoJSON, GML can be updated in any text editor. Each feature has a list of properties, geometry (points, lines, curves, surfaces and polygons) and spatial reference system.

There is generally more overhead when compare GML with GeoJSON. This is because GML results in more data for the same amount of information.

Google Keyhole Markup Language (KML/KMZ)

.KML.KMZ:

KML stands for Keyhole Markup Language. This GIS format is XML-based and is primarily used for Google Earth. KML was developed by Keyhole Inc which was later acquired by Google.

KMZ (KML-Zipped) replaced KML as being the default Google Earth geospatial format because it is a compressed version of the file. KML/KMZ became an international standard of the Open Geospatial Consortium in 2008.

The longitude, latitude components (decimal degrees) are as defined by the World Geodetic System of 1984 (WGS84). The vertical component (altitude) is measured in meters from the WGS84 EGM96 Geoid vertical datum.

GPS eXchange Format (GPX)

.GPX:

GPS Exchange format is an XML schema that describes waypoints, tracks and routes captured from a GPS receiver. Because GPX is an exchange format, you can openly transfer GPS data from one program to another based on its description properties.

The minimum requirement for GPX are latitude and longitude coordinates. In addition, GPX files optionally stores location properties including time, elevation and geoid height as tags.

IDRISI Vector

.VCT.VDC:

IDRISI vector data files have a VCT extension along with an associated vector documentation file with a VDC extension.

VCT format are limited to points, lines, polygons, text and photos. Upon the creation of an IDRISI vector file, it automatically creates a documentation file for building metadata.

Attributes are stored directly in the vector files. But you can optionally use independent data tables and value files.

MapInfo TAB

.TAB.DAT.ID.MAP.IND:

MapInfo TAB files are a proprietary format for Pitney Bowes MapInfo software. Similar to shape files, they require a set of files to represent geographic information and attributes.

- TAB files are ASCII format that link the associated ID, DAT, MAP and IND files.
- DAT files contain the tabular data associated as a dBase DBF file.
- ID files are index files that link graphical objects to database information.
- MAP files are the map objects that store geographic information.
- IND files are index files for the tabular data.

OpenStreetMap OSM XML

.OSM:

OSM files are the native file for OpenStreetMap which had become the largest crowdsourcing GIS data project in the world. These files are a collection of vector features from crowd-sourced contributions from the open community.

The GIS format OSM is OpenStreetMap's XML-based file format. The more efficient, smaller PBF Format ("Protocolbuffer Binary Format") is an alternative to the XML-based format.

The data interoperability in QGIS can load native OSM files. The OpenStreetMap plugin can convert PBF to OSM, which then can be used in QGIS.

Digital Line Graph (DLG)

.DLG:

Digital Line Graph (DLG) files are vector in nature that were generated on traditional paper topographic maps. For example, this includes township & ranges, contour lines, rivers, lakes, roads, railroads and towns.

Much of the U.S. Bureau of Census Topologically Integrated Geographic Encoding and Referencing (TIGER) data were generated using the standard DLG format.

Geographic Base File-Dual Independent Mask Encoding (GPF-DIME)

The GPF-DIME file format was developed by the US Census Bureau in the late 1960s as one of the first GIS data formats to exist. It was used to store the US road network for major urban areas, which is a key factor in census information.

GPF-DIME supports choropleth mapping and also assisted in removing error for digitizing features. DIME was a key component to the current TIGER (Topologically Integrated Geographic Encoding and Referencing) system, which was produced by the US Census Bureau.

ArcInfo Coverage

ArcInfo Coverages are a set of folders containing points, arcs, polygons or annotation. Tics are geographic control points and help define the extent of the coverage.

Attributes are stored in the ADF or INFOb tables. Each feature is identified with a unique number. These feature numbers are a way to link attribute data with each spatial feature.

Coverages were the standard format during the floppy disk era. But over time, this GIS format has become obsolete and mostly unsupported in GIS software.

Raster GIS File Formats

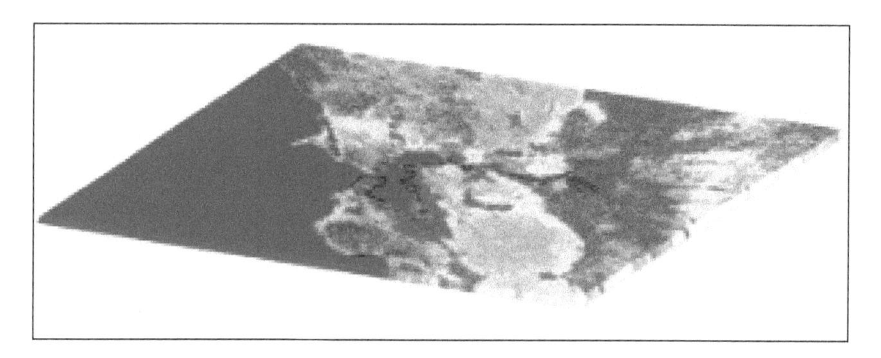

Raster data is made up of pixels (also referred to as grid cells). They are usually regularly-spaced and square but they don't have to be.

Rasters have pixel that are associated with a value (continuous) or class (discrete).

ERDAS Imagine (IMG)

.IMG:

ERDAS Imagine IMG files is a proprietary file format developed by Hexagon Geospatial. IMG files are commonly used for raster data to store single and multiple bands of satellite data.

IMG files use a hierarchical format (HFA) that are optional to store basic information about the file. For example, this can include file information, ground control points and sensor type.

Each raster layer as part of an IMG file contains information about its data values. For example, this includes projection, statistics, attributes, pyramids and whether or not it's a continuous or discrete type of raster.

American Standard Code for Information Interchange ASCII Grid

.ASC

ASCII uses a set of numbers (including floats) between 0 and 255 for information storage and processing. They also contain header information with a set of keywords.

In their native form, ASCII text files store GIS data in a delimited format. This could be comma, space or tab-delimited format. Going from non-spatial to spatial data, you can run a conversion process tool like ASCII to raster.

GeoTIFF

.TIF.TIFF.OVR:

The GeoTIFF has become an industry image standard file for GIS and satellite remote sensing applications. GeoTIFFs may be accompanied by other files:

- TFW is the world file that is required to give your raster geolocation.
- XML optionally accompany GeoTIFFs and are your metadata.
- AUX auxiliary files store projections and other information.
- OVR pyramid files improves performance for raster display.

IDRISI Raster

.RST.RDC:

IDRISI assigns RST extensions to all raster layers. They consist of numeric grid cell values as integers, real numbers, bytes and RGB24.

The raster documentation file (RDC) is a companion text file for RST files. They assign the number of columns and rows to RST files. Further to this, they record the file type, coordinate system, reference units and positional error.

Envi RAW Raster

.BIL.BIP.BSQ:

Band Interleaved files are a raster storage extension for single/multi-band aerial and satellite imagery.

Band Interleaved for Line (BIL) stores pixel information based on rows for all bands in an image.

Whereas Band interleaved by pixel (BIP) assigns pixel values for each band by rows.

Finally, Band sequential format (BSQ) stores separate bands by rows.

BIL files consist of a header file (HDR) that describes the number of columns, rows, bands, bit depth and layout in an image.

PCI Geomatics Database File (PCIDSK)

.PIX:

PIX files are raster storage layers developed by PCI Geomatics. It's a flexible file type that stores all image and auxiliary data called "segments" in a self-contained file. For example, segments can include image channels, training site and histogram information.

As a database file, PIX files can hold raster channels with varying bit depths. They can also store projections, attribute information, metadata and imagery/vectors.

Esri Grid

Grid files are a proprietary format developed by Esri. Grids have no extension and are unique because they can hold attribute data in a raster file. But the catch is that you can only add attributes to integer grids.

Attributes are stored in a value attribute tables (VAT) – one record for each unique value in the grid, and the count representing the number of cells.

The two types of Esri Grid files are integer and floating point grids. Land cover would be an example of a discrete grid. Each class has a unique integer cell value. Elevation data is an example of a floating point grid. Each cell represents an elevation floating value.

Compressed Raster File Formats

Lossy GIS compression reduces file size by permanently eliminating certain information, especially redundant information (even though the user may not notice it).

These lossy compression algorithms often result in greater reductions of file size. Here are examples of highly compressed GIS formats.

ER Mapper Enhanced Compression Wavelet

.ECW:

ECW is a compressed image format typically for aerial and satellite imagery. This GIS file type is known for its high compression ratios while still maintaining quality contrast in images.

ECW format was developed by ER Mapper, but it's now owned by Hexagon Geospatial.

Joint Photographic Experts Group JPEG2000

.JP2:

JPEG 2000 typically have a JP2 file extension. They are a wavelet compression with the latest JPG format giving an option for lossy or lossless compression.

JPEG 2000 GIS formats require a world file which gives your raster geolocation. They are an optimal choice for background imagery because of its lossy compression. JPEG 2000 can achieve a compression ratio of 20:1 which is similar to MrSID format.

LizardTech Multiresolution Seamless Image Database MrSID

.SID.SDW:

LizardTech's proprietary MrSID format is commonly used for orthoimages in need of compression. MrSID images have an extension of SID and are accompanied with a world file with the file extension SDW.

MrSIDs have impressive compression ratios. Color images can be compressed at a ratio of over 20:1. LizardTech's GeoExpress is the software package capable of reading and writing MrSID format.

Geographic Database File Formats

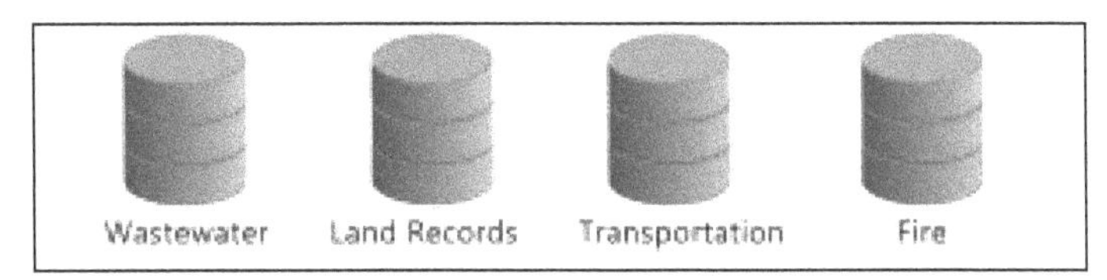

We store geographic data in various database file format. Databases are a structured set of data/information.

But the key difference is that geographic databases allow the storage of location information.

Esri File Geodatabase

.GDB:

Esri created the file geodatabase to be a container for storing multiple attribute tables, vector and raster data sets. It's the successor of the personal geodatabase (MDB) – and Esri recommends file geodatabases over personal geodatabases.

File geodatabases offer structural and performance advantages. They have fast performance, versatile relationships, compatible storage for rasters, improved spatial indexes, data compression, customizable configuration and 1 terabyte file size restrictions.

Within a geodatabase, geographic datasets are referred to as feature classes. But geodatabases can store more complex data such as networks, raster mosaics and feature data sets.

Esri Personal Geodatabase

.MDB:

Personal geodatabases use the default Microsoft Access database file extension (MDB). They used to be the most ubiquitous database type for managing geospatial data. Personal geodatabases were advantageous because you could manage multiple attribute tables, vector and raster datasets and create relationship classes.

But their biggest drawback was their limited 2GB in storage capacity. Whereas file geodatabases offer 2TB of capacity. In the end, you'd quickly reach storage capacity with personal geodatabases just by adding a couple of raster and vector data sets.

OGC GeoPackage

.GPKG:

GPKG are self-contained serverless SQLite databases that can contain anything from vector, tiles, rasters, layer attributes, and even extensions.

Unlike shapefile which have 3 mandatory files, this open standards geospatial container is easy to share because it's all contained in a single file.

Mapbox MBTiles

.MBTILES:

MBTILES are for storing and packaging sets of raster or vector map tiles in a single file. The file format is based on a SQLite database.

The only coordinate system MBTiles support is spherical Mercator. MBTiles file formats are designed for Mapbox and other web/mobiles applications.

GE Smallworld Version Managed Data Store

.VMDS:

Smallworld software is widely used in electrical, telecommunication, gas, water and utilities. It uses the VMDS "Version Managed Data Store" for database storage.

VMDS stores multiple types of raster and vector geometries in spatial and topological utility networks. They are also capable of querying and analysis in GE Smallworld.

SpatiaLite

.SL3,.SQLITE:

SpatiaLite uses the SQLite database engine. But SpatiaLite extends SQLite by giving it spatial capabilities.

SpatiaLite give similar functionality to geodatabases and are the rough equivalence to PostgreSQL + PostGIS. They are open source and lightweight with the ability to hold spatial and non-spatial files in a single file container.

Relational Database Management System (RDBMS) Enterprise

Relational Database Management Systems (RDBMS) are commonly used for multi-user editing environments.

They also support versioned editing, backups and recovery of an enterprise database over the same network.

PostGIS + PostgreSQL

Open source PostGIS adds spatial objects to the cross-platform PostgreSQL database. The three features that PostGIS delivers to PostgreSQL DBMS are spatial types, indexes and functions.

With support for different geometry types, the PostGIS spatial database allows querying and managing information about locations and mapping. PostGIS can be leveraged in several GIS software packages including QGIS, GRASS, ArcGIS and MapInfo.

ArcSDE Enterprise Geodatabase + (Oracle, Microsoft SQL Server, IBM DB2)

Relational Database Management Systems (RDBMS) and ArcSDE support versioned editing, backups and recovery with multiple users over the same network. DBMS storage models include Oracle, Microsoft SQL Server, IBM DB2/Informix and PostgreSQL.

ArcSDE serves data in a centralized way over an entire organization using a relational database management system. End-users can access spatial data in an Esri environment and seamlessly edit and analyze data in an enterprise geodatabase.

LiDAR File Formats

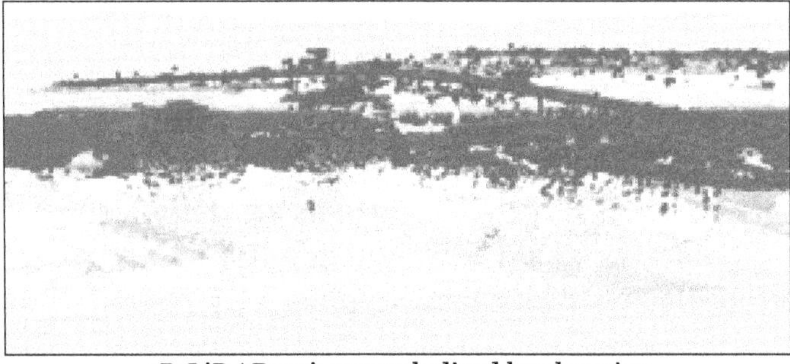

3D LiDAR points symbolized by elevation.

The growth of Light Detection and Ranging (LiDAR) technology has revolutionized how we view the surface of Earth.

As point cloud data, LiDAR is a dense network of coordinate points with elevation values. These GIS formats require specialized software or extensions to view or edit.

ASPRS LiDAR Data Exchange Format

.LAS,.LASD,.LAZ:

The LAS file format is a binary file format specifically for the interchange between vendors and customers. Overall, LAS files maintain information specific to LIDAR without the loss of information.

LAS files are available for public use, unlike ASCII and other proprietary file formats. The dense networks of coordinate point measurements are so large sometimes that they often need to be split to prevent the file size becoming too large.

When you compress a LAS file, the file format specifically for this is LAZ. You can save significant storage space using the LAZ file format. Like most file compression, LAZ has no information loss.

Lastly, LAS Datasets (LASD) reference a set of LAS files. The purpose of LASD is to be able to examine 3D point cloud properties from the referenced LAS files. Through LAS datasets, you can visualize triangulated surfaces and perform statistical analysis.

Point Cloud XYZ

.XYZ:

XYZ files don't have specifications for storing point cloud data. The first 3 columns generally represent X, Y and Z coordinates. But there's no standard specification so it may include RGB, intensity values and other LiDAR values.

They are in the ASCII point cloud group of file formats which includes TXT, ASC and PTS. Non-binary files like XYZ are advantageous because they can be opened and edited in a text editor.

CAD File Formats

CAD (computer assisted drafting) often goes hand-in-hand with Geographic Information System (GIS). GIS imports design models that were likely built in Autodesk or Bentley Systems MicroStation.

Autodesk Drawing

.DWF.DWG,.DXF:

Autodesk CAD (computer assisted drafting) file formats are designed for 2D and 3D designs. They generally contain elements such as edges, curves, annotation text in layers. DWG/DXF are vector files that use Cartesian coordinates. Every element plots XY points in a grid.

- DWF (Design Web Format) is more specific for view and use on the internet.

- DWG (DraWinG) is the native format and working version for AutoCAD containing metadata.
- Lastly, DXF (Drawing Exchange Format) stores drawing information as exact representations of the data. But the purpose of DXF was for data exchange between CAD programs.

Bentley Microsystems DGN File Format

.DGN:

DGN or "Design" is the native format for Bentley Systems MicroStation. Similar to other CAD design formats, engineers and architects use it for construction design.

DGN files consist of layers including annotation, points, polylines, polygons and multipath. They also contain style information (ColorIndex) and a spatial reference system.

Elevation File Formats

Elevation file formats are specific to digital elevation model products. For example, the USGS DEM and Canadian CDED capture regularly-spaced elevation values in a raster grid.

USGS DEM, Canadian CDED

.DEM:

The DEM format are raster-based ASCII files specifically developed by the USGS to capture digital elevation models. They are widely used in the industry because of the high volume of legacy elevation models produced by the USGS.

The DEM format is a single file containing 3 record types:

- Record A stores general characteristics of the DEM such as descriptive name, elevation minimum and maximum, extent boundaries and number of B records.
- Record B contains a header and elevation profile.
- Record C stores the accuracy of the data and is optional.

Digital Terrain Elevation Data (DTED)

.DT0.DT1.DT2:

Digital Terrain Elevation Data (DTED) is a standard format created by the National Geospatial-Intelligence Agency. They are a raster format consisting of terrain elevation values often captured from aircraft radar. User-defined attributes are assigned through TAB files.

The 3 levels of resolutions contain various cell-spacing resolution:

- Level 0 spacing is 30 arc second spacing (nominally one kilometer).
- Level 1 spacing is 3 arc seconds (approximately 100 meters).
- Level 2 spacing is 1 arc second (approximately 30 meters).

Web File Formats

These web file formats are built specifically to serve and display geographic features over the internet.

Although there are other web-based file formats that stores geographic data (such as GeoJSON), these file formats are unique to web mapping.

GeoRSS

URL.XML:

Websites publish RSS feeds to subscribers, which are provided in a XML file. GeoRSS extends XML feeds to include geographic data. The GeoRSS specification describes how to add spatial geometries like points, lines and polygons to XML feeds.

Webfeeds with location have become a tool for disaster notification. For example, the USGS publishes real-time earthquakes through GeoRSS. Now, RSS have locations.

Web Feature Service (WFS)

URL:

Web feature services allows users to share geospatial (or non-spatial) over the internet. Thus, feature services can be consumed through the internet in web maps, desktop and web applications. Users access web features services by pointing to the REST endpoints or URL.

- Web Feature Services (WFS) are for editing features.
- Web Mapping Services (WMS) display features without editing capabilities.

Esri ArcGIS Online Web Services

URL:

ArcGIS Online is Esri's cloud-based platform that allows users to publish content online to share collaboratively with other organizations or the general public.

- Feature layers are vector layers that can be viewed and edited by users in an organization.
- Tile layers are pre-drawn commonly used for base maps.
- Scene layers are specific for a collection of three-dimensional data.

Multitemporal File Formats

Temporal data has a time component attached to it. A lot of weather data uses temporal GIS data formats because how important time is related to weather.

Other examples of temporal data are demographic trends, land use patterns and lightning strikes.

Network Common Data Form (NetCDF)

.NC:

NetCDF GIS format is an interface for array-oriented data for storing multi-dimensional variables. An example of a multi-dimension NetCDF could be temperature, precipitation or wind speed over time. It's commonly used for scientific data involved in the oceanic and atmospheric community as a GIS data storage format.

The ArcGIS multidimensional toolbox and the QGIS NetCDF Browser both support NetCDF files.

Hierarchical Data Format

.HDF:

HDF (Hierarchical Data Format) was designed by the National Center for Supercomputing Applications (NCSA) to manage extremely large and complex scientific data. It's a versatile data model with no limit on the number or size of data objects in the collection.

ArcGIS is capable of reading HDF4 and HDF5 data. The free open source GDAL (command-line) tools supports the conversion of HDF files to GeoTIFF. The HDF View program allows users to view HDF files.

GRIdded Binary or General Regularly-distributed Information in Binary (GRIB)

.GRIB:

Similar to NetCDF, GRIB files are commonly used in meteorology to store historical and forecast weather data. It's a multidimensional file with the advantages of self-description, flexibility and expandability.

GRIB is standardized by the World Meteorological Organization's Commission and in operation since 1985. Currently, there are three versions of GRIB files (GRIB 0, 1 and 2). There are tools to convert GRIB into rasters such as grb2grid and QGIS software.

GIS Software Project File Formats

GIS project files are used in GIS applications. Generally, they all store layers in a hierarchical manner and then display them in a layout.

They retain symbology, queries, labeling and other properties for building maps.

Map Exchange Document (MXD)

.MXD:

MXD stands for Map eXplorer Document. ArcGIS uses this file format to store map layers in a table of contents. Each layer in a data frame references a data source.

Map layers are displayed from the map layout in a hierarchical manner. When reopening a MXD, all symbology and labeling are retained since it was last saved

QGIS 2.X Project File

.QGS:

The QGS extension is a project file for the GIS software program QGIS (formerly Quantum GIS). This file type can be opened similar to .TXT or .XLS file.

All the map layers and composers are stored in a QGS project file. It retains the same, labeling and map layers as they were since last saving. Map layers are referenced pointing to the physical data sources.

QGIS projects generate a backup of the project file automatically with the file extension QGS~. These files are stored in the same directory as the project file.

ArcGIS Pro Project File

.APRX:

ArcGIS Pro project files (APRX) contain maps, toolboxes, databases, folders and even styles. They can also contain connections to databases, servers and folders.

APRX files are the successor to MXDs, which were the equivalent to project files in Esri ArcGIS. But they are different from MXDs in that projects can have multiple maps and layouts in a single project.

QGIS 3 Project File

.QGZ:

QGZ is the default project file for QGIS 3.2 and greater. This zipped container stores the QGS XML file and is used for storing QGIS layouts, properties and layers.

Map Exchange Document Template (MXT)

.MXT:

Map Exchange Document Template (MXT) are standardized layouts for Esri ArcGIS. They contain common basemaps and page layouts to be reused repeatedly in an organization.

Your ArcGIS profile uses the normal.mxt. In order to fix map document issues, you can reset your application through the normal.mxt.

MapInfo Workspace

.WOR:

.MWS Map composition files (.MAP) store a set of map layers, symbology and color palettes in a file with a MAP extension. Once you reopen it, MAP files recreate the map layout as needed for printing.

Esri ArcGlobe Document

.3DD:

ArcGlobe is a global 3D visualization and analysis environment focusing on larger study areas. 3DD is the extension for ArcGlobe which houses all your feature and raster layers in a global view.

Esri ArcScene Document

.SXD:

ArcScene is a 3D feature and raster viewer specializing in smaller study area or local scenes. SXD is the extension for ArcScene that saves the scenes view, layers and properties.

IDRISI Map Composition File

.MAP:

Map composition files (.MAP) store a set of map layers, symbology and color palettes in a file with a MAP extension. Once you reopen it, MAP files recreate the map layout as needed for printing.

Cartographic File Formats

The purpose of cartographic file formats is to standardize map creation with a set of symbols, labels or feature display.

Generally, they don't hold any of the physical data. But they contain the symbology to stylize your map features.

Esri ArcGIS Layer File

.LYR.LYRX:

Layer files are used for displaying a set of symbology in a map. They don't contain the geographic data itself. Instead layer files simply specify how the data will be displayed.

When you share a vector or raster data set, a layer file ensures the same symbology will be displayed in another map.

- LYR files are the extension for Esri ArcGIS 10.X.
- LYRX files are for ArcGIS Pro.

QGIS Layer Definition File

.QLR:

Not only do QLR files contain the styling information for a layer, they also point to a referenced layer. QLR files don't hold any of the physical data because their purpose is for storing symbology and queries. QLR files are XML based so you can freely open and edit them in a text editor.

Esri ArcGIS Style File

.STYL.STYLX:

STYL files are a set of symbols that can be assigned to symbolize features in a map layout. They often carry a list of icons specific to a theme such as forestry, petroleum or geology. Once you add a STYL file to a map document, map features can obtain any unique symbology as part of the style file.

- STYLX extension are style files for ArcGIS Pro.
- STYL files are style files for Esri ArcGIS 10.

QGIS Style File

.QML:

QGIS layer style files (*.QML) contain the symbology and labelling to style how features are viewed in a QGIS project.

You can apply a QML file to any file without needing data. Because QML is in an XML format, you can open and edit it in a text editor.

3D File Formats

Three-dimensional file formats not only give XY locations of features, but they also add depth to features.

These 3D file formats are graphic representations of objects in the real world developed in 3D modelling software.

COLLADA

.DAE:

COLLADA are 3D object representations stored in a XML file. This reference image file simulates textures in 3D web scenes in Esri and Google Earth.

COLLADA is a format that you can transfer from 3D modelling software such as Trimble Sketch-Up, Autodesk Maya and 3DS Max.

Trimble Sketchup

.SKP:

Sketchup files (SKP) are 3D object representations native to the three-dimensional modelling software Trimble Sketchup.

These conceptual design files (buildings, towers, trees, etc) can be placed in Google Earth.

Interchange File Formats

The purpose of interchange files are to transfer files between different software systems.

Generally, they are non-native formats specifically designed for interoperability and data transfer.

Esri ArcInfo Export (Interchange)

.E00.E01.E02:

Esri ArcInfo Interchange files are no longer supported. This obsolete GIS format was used to conveniently exchange GIS coverage files.

It has the extension E00 and increases incrementally (E01, E02...) with individual coverage files. Although convenient for interchange, you need to process the data before you can add it to ArcGIS.

In order to convert Esri ArcInfo Interchange files, you can run the 'Import from E00' tool in the Conversion Arc Toolbox.

MapInfo Interchange File

.MIF.MID:

MIF format are versatile files that allows the exchange of MapInfo files between different geospatial systems. This ASCII file contains two files:

- MIF stores the graphics.
- MID retains the attributes as delimited text.

Map Package

.MPK:

Map package (MPK) files are an Esri storage container that contains all the elements and underlying data of a MXD. The purpose of generating MPKs is to not only transfer the layers in a table of contents, but the physical data that is associated to each layer in a data frame.

It's common to exchange map packages to users outside an organization who do not have access to the network data sources. Once the MPK file is transferred, they have access to editing their own source version of data

Other GIS File Formats

The remaining GIS formats don't belong in any specific group. They are geographic in nature and perform a specific function related to the analysis, management or display of geographic information.

Esri ArcGIS Toolbox:

.TBX:

TBX extensions carry the geoprocessing tools that you can analyze geospatial data in Esri

ArcGIS 10.X. When you create a custom toolbox, you can add toolsets and models. For example, you can add Python scripts (.PY) and models from ModelBuider for specific routines and processes.

ArcSDE Connection File

.SDE:

SDE files connect to a SDE geodatabase including the username, password, version and database path. The purpose of ArcSDE connection files are to gain access to an instance of your enterprise geodatabase on a network. You authenticate the database from either operating system or database authentication

Adobe Geospatial PDF

.PDF:

Geospatial PDFs store points, lines, polygons and raster layers all represented in geographic space. By using this file format, users can measure distance, adjust coordinate systems as well as add edit locations. Geospatial PDFs can retain tabular data associated with each graphic.

Apple Venue Format

.AVF:

Apple Venue Format (AVF) is the standard format for Apple's indoor maps. AVF stores GeoJSON files in a dataset folder. Each GeoJSON file represents a feature as part of a data collection.

GPS Exchange Format

GPX, or GPS Exchange Format, is an XML schema designed as a common GPS data format for software applications. It can be used to describe waypoints, tracks, and routes. The format is open and can be used without the need to pay license fees. Location data (and optionally elevation, time, and other information) is stored in tags and can be interchanged between GPS devices and software. Common software applications for the data include viewing tracks projected onto various map sources, annotating maps, and geotagging photographs based on the time they were taken.

Data Types

These are the essential data contained in GPX files. Ellipsis means that the previous element can be repeated. Additional data may exist within every markup but is not shown here:

- wptType is an individual waypoint among a collection of points with no sequential relationship. It consists of the WGS 84 (GPS) coordinates of a point and possibly other descriptive information.
- rteType is a route, an ordered list of routepoint (waypoints representing a series of significant turn or stage points) leading to a destination.

- trkType is a track, made of at least one segment containing waypoints, that is, an ordered list of points describing a path. A Track Segment holds a list of Track Points which are logically connected in order. To represent a single GPS track where GPS reception was lost, or the GPS receiver was turned off, start a new Track Segment for each continuous span of track data.

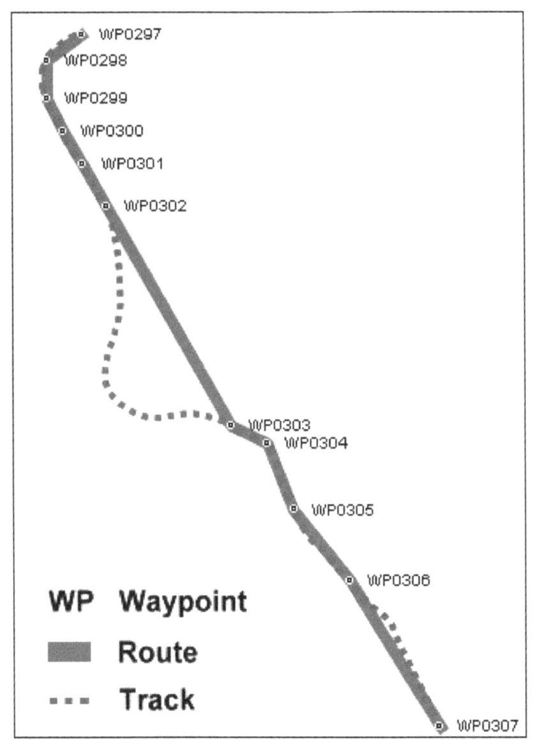

Waypoints, routes and tracks recorded by GPS receivers.

Conceptually, tracks are a record of where a person has been and routes are suggestions about where they might go in the future. For example, each point in a track may have a timestamp (because someone recorded where *and when* they were there), but the points in a route are unlikely to have timestamps (other than estimated trip duration) because route is a suggestion which might never have been traveled.

Technically:

- A track is made of a sufficient number of trackpoints to precisely draw every bend of a path on a bitmap. It is:
    - The raw output of, for example, a GPS recording the course of one's trip,
    - The rearrangement of such points to remove excess ones,
    - Data from any source such as extracted from a vector map.
- A route is made of routepoints between which a program must get the corresponding track from a vector map to draw it. The routepoints may be crossings or junctions or as distant as stopover towns, such as those making a trip project. Hence, such a project can be saved and reloaded in a GPX file.

- A process called routing computes a route and may produce a GPX route made of the route-points where some driver action takes place (turn left, enter a roundabout, limit speed, name of the next direction road, etc.). The GPX points may contain the text of those instructions.
- The GPX file may contain both route and track so that a program can get precise points from the track even if it has no access to a vector map.

The minimum properties for a GPX file are latitude and longitude for every single point. All other elements are optional. Some vendors, such as Humminbird and Garmin, use extensions to the GPX format for recording street address, phone number, business category, air temperature, depth of water, and other parameters.

Units

Latitude and longitude are expressed in decimal degrees, and elevation in meters, both using the WGS 84 datum. Dates and times are not local time, but instead are Coordinated Universal Time (UTC) using ISO 8601 format.

Spatial Archive and Interchange Format

The Spatial Archive and Interchange Format was defined in the early 1990s as a self-describing, extensible format designed to support interoperability and storage of geospatial data.

SAIF Dataset

SAIF has two major components that together define SAIFtalk. The first is the Class Syntax Notation (CSN), a data definition language used to define a dataset's schema. The second is the Object Syntax Notation (OSN), a data language used to represent the object data adhering to the schema. The CSN and OSN are contained in the same physical file, along with a directory at the beginning of the file. The use of ASCII text and a straightforward syntax for both CSN and OSN ensure that they can be parsed easily and understood directly by users and developers. A SAIF dataset, with a. saf or. zip extension, is compressed using the zip archive format.

Schema Definition

SAIF defines 285 classes (including enumerations) in the Class Syntax Notation, covering the definitions of high-level features, geometric types, topological relationships, temporal coordinates and relationships, geodetic coordinate system components and metadata. These can be considered as forming a base schema. Using CSN, a user defines a new schema to describe the features in a given dataset. The classes belonging to the new schema are defined in CSN as subclasses of existing SAIF classes or as new enumerations.

A ForestStand::MySchema for example could be defined with attributes including age, species, etc. and with ForestStand::MySchema specified as a subclass of GeographicObject, a feature defined in the SAIF standard. All user defined classes must belong to a schema, one defined by the user or previously existing. Different schemas can exist in the same dataset and objects defined under one schema can reference those specified in another.

Inheritance

SAIF supports multiple inheritance, although common usage involved single inheritance only.

Object Referencing

Object referencing can be used as a means of breaking up large monolithic structures. More significantly, it can allow objects to be defined only once and then referenced any number of times. A section of the geometry of the land-water interface could define part of a coastline as well as part of a municipal boundary and part of a marine park boundary. This geometric feature can be defined and given an object reference, which is then used when the geometry of the coastline, municipality and marine park are specified.

Multimedia

Multimedia objects can also be objects in a SAIF dataset and referenced accordingly. For example, image and sound files associated with a given location could be included.

Model Transformations and Related Software Applications

The primary advantage of SAIF was that it was inherently extensible following object oriented principles. This meant that data transfers from one GIS environment to another did not need to follow the lowest common denominator between the two systems. Instead, data could be extracted from a dataset defined by the first GIS, transformed into an intermediary, i.e., the semantically rich SAIF model, and from there transformed into a model and format applicable to the second GIS.

This notion of model to model transformation was deemed to be realistic only with an object oriented approach. It was recognized that scripts to carry out such transformations could in fact add information content. When Safe Software developed the Feature Manipulation Engine (FME), it was in large measure with the express purpose of supporting such transformations. The FMEBC was a freely available software application that supported a wide range of transformations using SAIF as the hub. The FME was developed as a commercial offering in which the intermediary could be held in memory instead of as a SAIF dataset.

Applied Spatial Intelligence

The foundation of spatial intelligence is knowing "where" information happens and the effect it has on its surroundings. Applied Spatial Intelligence can be defined as "the practice of using spatial technologies such as Geographic Information Systems (GIS), Aerial and Satellite Imagery, Global Navigation Satellite Systems (GPS, Glonass, Galileo) and Spatial Databases to visualize and enhance your understanding of natural occurrences, human impact and situational or hypothetical conditions".

Today with the proliferation of cloud computing and mobile devices and the explosion of web maps, global technology leaders such as Esri, Google, Microsoft and Apple are pushing the boundaries of spatial awareness. The benefits derived from location-based information are now known to save lives by preventing crime and the spread of disease, conserve resources by increasing efficiencies and profitability, control pollution by identifying and tracking illicit discharges... the list goes on and on. It doesn't matter who you are, where you are or what you do, Applied Spatial Intelligence is mission critical.

GEOSPATIAL INTELLIGENCE

Geospatial intelligence is a new subcategory of GIS science, including technology, critical information about an issue and a deep analysis of the field we study. The advantage of geospatial intelligence is that gives direct results for critical issues like strategic defense at a political and national level, business and humanitarian intelligence, etc.

In simple words, GEOINT (GEOspatial INTelligence) refers to the science that analyzes imagery from satellites and aircraft to describe, assess and depict, natural and anthropogenic features, and georeferenced activities. Geospatial intelligence analysis includes analysis of UV and microwave data of electromagnetic spectrum, georeferenced social media, global positioning data, analysis of spatial, radiometric, temporal and history data. The list doesn't stop here but it is too long to be mentioned as a whole! To conclude, the means by which the above data are transmitted, are usually stationary and moving.

Stations by electro-optical sensor programs (satellites/ aircraft) such as IR, MWIR, SWIR, TIR. Another source is GPS and on-site data by ground instruments.

Geospatial intelligence started from US military in order to take rapid decisions about issues of national defense. Nowadays, there is wide interest in GEOINT, it is progressing sharply and in different ways depending on the governments, while other countries such as India are holding tight the conferences of geospatial intelligence.

It is obvious, that geospatial intelligence is a rapidly growing industry, at the forefront of interest, for many countries. Moreover, since we talk about "geospatial" there is no doubt about its correlation with GIS.

Principles Govern Geospatial Intelligence

It is widely known that GIS supports a big variety of applications targeted to geospatial information. It is reasonable to wonder which is the difference of geospatial intelligence with all the other applications, from the moment that everything in GIS rotates around geospatial data and the correlation with general data. Well, this question answered by Bacastow, who enacted some rules and principles to make *geospatial intelligence* fully understandable by everyone, as well as to emphasize to its differentness.

First of all, GEOINT seems to focus on the roots of geospatial science, technology and thinking in order to achieve a better decision making. Moreover, the geospatial analysis happens as a logical sequence of natural human entities and technical events. Bacastow mentions that the natural landscape as well as the human awareness about the earth, affect and restrict the human action about any issues. To conclude with the uniqueness of geospatial intelligence another conclusion of Bacastow was that GEOINT pursues to find out correlations between geospatial data and patterns of life, perceptions, customs and prejudices over the time, not only by data mapping but also through the technical systems created and used by geospatial intelligence analysts, who encounter the science according to their acuities and discernment.

Correlation of Geospatial Intelligence with GIS

Let's start with the basics of GIS program. It is a system that integrates a whole software inside it, as well as hardware and general data in order to manage, analyze, and depict them in geospatial level. Furthermore, GIS program gives the ability for the user to view the results visualized and understand them in depth, employ them and extract relationships between the correlated data. In that way, it is easier to deepen in a problem and solve it.

After that, we can realize that geospatial intelligence is a science, or a technology and GIS program is a tool or software which help us apply and develop GEOINT. Of course, there is much software where we can apply geospatial intelligence, but GIS is the perfect one because it gives to the user a wide variety of advantages such as the ability for multilayering with geology, contour, elevation, topographic and other maps. GIS program supports also satellite imagery. Moreover, we can easily transform conventional data in a digital form by hand-tracing digitization or use a digitizing tablet which helps us collect and upload coordinates. Scanners and the global positioning system are some other tools to collect and digitize data. Tabular data, Excel files and other files from different software compatible with GIS can also get uploaded and utilized in GIS. Last but not least, we have the opportunity to upload to GIS program previously digitized data or transship data to WebGIS from other online sources.

Benefits of Geospatial Intelligence through GIS Program

Geospatial intelligence can't stand alone. It is necessary to have additional information and the contribution of GIS to draw conclusions for an issue. However, finding further information namely data is easy anymore. Having access to GIS is also easy through WebGIS. All the disadvantages are eliminated one by one over the time, while the benefits are increasing:

- Better management of various concepts, from the right promotion of a cosmetic to the confrontation of an epidemic.
- Easy decision making with maps including many parameters.
- Improvement of productivity with better planning of the campaign, better research for the right place to locate a shop etc.
- Great provision and help to human actions
- Creation of common route for organizations or companies in order to share information and data and evolve science and each profession.
- Visualized maps of people movements, land cover, and terrain planning.
- Direct results and decision making in emergency situation (national defense, problem in nuclear reactor, extreme weather phenomena etc).

It is obvious that geospatial intelligence is a science targeted to more profound concepts and that is why it is used for national defense and other crucial programs. Moreover, it combines knowledge from many different disciplines such as UAV and IT technologies, database design and management, software development, imagery analysis, and the most important knowledge and critical thinking about geography, human, environment, and ethics.

Uses of Geospatial Intelligence

- The role of machine learning and GEOINT in disaster response.
- Open geospatial data platforms and food scarcity.
- Interoperability for GEOINT applications and data in the military.
- The role of data stewardship in crisis mapping.

One of the most significant trends in geospatial intelligence is the shift in creation and ownership of data. As the United States Geospatial Intelligence Foundation noted, new data sources like Open Street Map and geotagged social media pictures can be leveraged for vital intelligence. However, the availability and open nature of these platforms also presents challenges for the GEOINT community, which must rely on data it no longer has full ownership and control over.

Machine Learning and GEOINT: Managing the Chaos of Natural Disasters

While it may sometimes seem like the entire world is documented, catalogued and analyzed, there are still many permanent and semi-permanent structures that remain unmapped. One of the main barriers to collecting geospatial data has been the manual and time-intensive work involved; this is especially problematic for instances where landscapes and structures change dramatically (i.e. after a natural disaster).

Geospatial intelligence software, augmented with machine learning, could help to map changes in terrain and structures, making disaster response projects more efficient and more effective. Several organizations are looking toward algorithms to help create more timely and accurate maps. One example is the SpaceNet "Road Detection and Routing Challenge," a $50,000 competition to develop an automated method for extracting information about road networks. Crowdsourced data proved to be an invaluable resource in the response to Hurricane Maria in Puerto Rico, but the successful implementation of machine learning could yield faster and more accurate maps to help emergency personnel find people in need or identify the best routes for delivering supplies.

Open Geospatial Data Platforms Helping Fight World Hunger

One of the core challenges in hunger worldwide is the fact that scarcity situations have usually already become dire by the time humanitarian efforts can begin. This is just one of the major challenges that DARPA is hoping to solve through a $7.2 million project awarded to Descartes Labs. The company hopes to create a vast geospatial data repository, leveraging sensors, satellite imagery and data from 75 different partners.

In 2017, the company hosted a hackathon, giving developers the goal of addressing food security issues. The resulting projects included:

- A platform to help farmers share information that would help regions protect crop yields and prevent scarcity.

- The development of a food security risk index.
- A fish distribution system for optimizing delivery to regions affected by drought.

Beyond the direct impact of reducing suffering from food shortages, the company suggests that addressing scarcity before it becomes a dire problem could help to avoid conflicts over resources.

Interoperability Drives the Future of Joint GEOINT Operations

The U.S. military has been a long-standing user of GIS intelligence to resolve conflicts, protect troops, assess risks and gain information about enemy operations. While not a new trend, the military has addressed new challenges.

One of the most important shifts in the way the military uses geospatial intelligence was the adoption of the object-based production framework. This philosophy focuses GEOINT around assembling data together around specific issues, rather than tasking analysts with collecting information from many different sources. This way, analysts spend more time developing intelligence and insights rather than with data management.

This approach is especially valuable in multinational joint operations, where data and GIS applications must be interoperable to ensure all stakeholders have access to mission-critical information.

Geospatial Data Stewardship as a Critical Factor in Improving Crisis Mapping

Although the visualizations and analyses provided to emergency responders have drastically improved our ability to respond to events likes hurricanes and other natural disasters, it is just one factor in how GEOINT has evolved. During the response to Hurricane Maria, for example, geospatial data was plentiful but disparate and difficult to use. This led to problems like duplicate deliveries and deliveries that were scheduled, but never made.

One of the developments to arise out of problems like these has been a rise in self-service geospatial intelligence products. For example, FEMA's GeoPlatform Disasters Portal provides curated geospatial information and datasets from numerous other apps and sources, providing a key data stewardship role. This effectively gives first responders and GEOINT teams a running start in responding to natural disasters.

One of the core themes in all the above GEOINT uses is the vast volume of data. As we look toward the future, the ability to manage data at large volumes will continue to be a key theme. However, it's important to note that the GEOINT industry will require expertise both in the analysis and in the preparation of that data.

ERRORS IN SPATIAL ANALYSIS

Errors can be injected at many points in a GIS analysis, and one of the largest sources of error is the data collected. Each time a new dataset is used in a GIS analysis, new error possibilities are also introduced. One of the feature benefits of GIS is the ability to use information from many sources, so the need to have an understanding of the quality of the data is extremely important.

Accuracy in GIS is the degree to which information on a map matches real-world values. It is an issue that pertains both to the quality of the data collected and the number of errors contained in a dataset or a map. One everyday example of this sort of error would be if an online advertisement showed a sweater of a certain color and pattern, yet when you received it, the color was slightly off.

Precision refers to the level of measurement and exactness of description in a GIS database. Map precision is similar to decimal precision. Precise location data may measure position to a fraction of a unit (meters, feet, inches, etc.). Precision attribute information may specify the characteristics of features in great detail. As an example of precision, say you try on two pairs of shoes of the same size but different colors. One pair fits as you would expect, but the other pair is too short. Do you suspect a quality issue with the shoes or do you buy the shoes that fit? Would you do the same when selecting GIS data for a project?

The more accurate and precise the data, the higher cost to obtain and store it because it can be very difficult to obtain and will require larger data files. For example, a 1-meter-resolution aerial photograph will cost more to collect (increased equipment resolution) and cost more to store (greater pixel volume) than a 30-meter-resolution aerial photograph.

Highly precise data does not necessarily correlate to highly accurate data nor does highly accurate data imply high precision data. They are two separate and distinct measurements. Relative accuracy and precision, and the inherent error of both precision and accuracy of GIS data determine data quality.

Difference between accuracy and precision:

Accuracy is the degree to which information on a map or in a digital database matches true or accepted values. Accuracy is an issue pertaining to the quality of data and the number of errors contained in a dataset or map. it is possible to consider horizontal and vertical accuracy with respect to geographic position, as well as attribute, conceptual, and logical accuracy.

- The level of accuracy required for particular applications varies greatly.
- Highly accurate data can be very difficult and costly to produce and compile.

Precision refers to the level of measurement and exactness of description in a GIS database. Precise locational data may measure position to a fraction of a unit. Precise attribute information may specify the characteristics of features in great detail. It is important to realize, however, that precise data – no matter how carefully measured – may be inaccurate. Surveyors may make mistakes or data may be entered into the database incorrectly.

- The level of precision required for particular applications varies greatly. Engineering projects such as road and utility construction require very precise information measured to the millimeter or tenth of an inch. Demographic analyses of marketing or electoral trends can often make do with less, say to the closest zip code or precinct boundary.
- Highly precise data can be very difficult and costly to collect. Carefully surveyed locations needed by utility companies to record the locations of pumps, wires, pipes and transformers to collect.

High precision does not indicate high accuracy nor does high accuracy imply high precision. But high accuracy and high precision are both expensive.

GIS practitioners are not always consistent in their use of these terms. Sometimes the terms are used almost interchangeably and this should be guarded against.

Two additional terms are used as well:

- Data quality refers to the relative accuracy and precision of a particular GIS database. These facts are often documented in data quality reports.
- Error encompasses both the imprecision of data and its inaccuracies.

Types of Error

Positional error is often of great concern in GIS, but error can actually affect many different characteristics of the information stored in a database.

Positional Accuracy and Precision

This applies to both horizontal and vertical positions.

Accuracy and precision are a function of the scale at which a map (paper or digital) was created. The mapping standards employed by the United States Geological Survey specify that:

"requirements for meeting horizontal accuracy as 90 percent of all measurable points must be within 1/30th of an inch for maps at a scale of 1:20,000 or larger, and 1/50th of an inch for maps at scales smaller than 1:20,000."

Accuracy Standards for Various Scale Maps:

1:2,400 ± 6.67 feet

1:4,800 ± 13.33 feet

1:10,000 ± 27.78 feet

1:12,000 ± 33.33 feet

1:24,000 ± 40.00 feet

1:63,360 ± 105.60 feet

1:100,000 ± 166.67 feet

1:1,200 ± 3.33 feet

This means that when we see a point on a map we have its "probable" location within a certain area. The same applies to lines.

Beware of the dangers of false accuracy and false precision, that is reading locational information from map to levels of accuracy and precision beyond which they were created. This is a very great danger in computer systems that allow users to pan and zoom at will to an infinite number of scales. Accuracy and precision are tied to the original map scale and do not change even if the user zooms in and out. Zooming in and out can, however, mislead the user into believing – falsely – that the accuracy and precision have improved.

Attribute Accuracy and Precision

The non-spatial data linked to location may also be inaccurate or imprecise. Inaccuracies may result from mistakes of many sorts. Non-spatial data can also vary greatly in precision. Precise attribute information describes phenomena in great detail. For example, a precise description of a person living at a particular address might include gender, age, income, occupation, level of education, and many other characteristics. An imprecise description might include just income, or just gender.

Conceptual Accuracy and Precision

GIS depend upon the abstraction and classification of real-world phenomena. The users determines what amount of information is used and how it is classified into appropriate categories. Sometimes users may use inappropriate categories or misclassify information. For example, classifying cities by voting behavior would probably be an ineffective way to study fertility patterns. Failing to classify power lines by voltage would limit the effectiveness of a GIS designed to manage an electric utilities infrastructure. Even if the correct categories are employed, data may be misclassified. A study of drainage systems may involve classifying streams and rivers by "order," that is where a particular drainage channel fits within the overall tributary network. Individual channels may be misclassified if tributaries are miscounted. Yet, some studies might not require such a precise categorization of stream order at all. All they may need is the location and names of all stream and rivers, regardless of order.

Logical Accuracy and Precision

Information stored in a database can be employed illogically. For example, permission might be given to build a residential subdivision on a floodplain unless the user compares the proposed plat with floodplain maps. Then again, building may be possible on some portions of a floodplain but the user will not know unless variations in flood potential have also been recorded and are used in the comparison. The point is that information stored in a GIS database must be used and compared carefully if it is to yield useful results. GIS systems are typically unable to warn the user if inappropriate comparisons are being made or if data are being used incorrectly. Some rules for use can be incorporated in GIS designed as "expert systems," but developers still need to make sure that the rules employed match the characteristics of the real-world phenomena they are modeling.

Finally, It would be a mistake to believe that highly accurate and highly precision information is needed for every GIS application. The need for accuracy and precision will vary radically depending on the type of information coded and the level of measurement needed for a particular application. The user must determine what will work. Excessive accuracy and precision is not only costly but can cause considerable error in details.

Sources of Inaccuracy and Imprecision

There are many sources of error that may affect the quality of a GIS dataset. Some are quite obvious, but others can be difficult to discern. Few of these will be automatically identified by the GIS itself. It is the user's responsibility to prevent them. Particular care should be devoted to checking for errors because GIS are quite capable of lulling the user into a false sense of accuracy and precision unwarranted by the data available. For example, smooth changes in boundaries, contour lines, and the stepped changes of chloropleth maps are "elegant misrepresentations" of reality. In fact, these features are often "vague, gradual, or fuzzy" (Burrough 1986). There is an inherent imprecision in cartography that begins with the projection process and its necessary distortion of some of the data (Koeln and others 1994), an imprecision that may continue throughout the GIS process. Recognition of error and importantly what level of error is tolerable and affordable must be acknowledged and accounted for by GIS users.

Burrough (1986) divides sources of error into three main categories:

1. Obvious sources of error.
2. Errors resulting from natural variations or from original measurements.
3. Errors arising through processing.

Generally errors of the first two types are easier to detect than those of the third because errors arising through processing can be quite subtle and may be difficult to identify. Burrough further divided these main groups into several subcategories.

Obvious Sources of Error

Age of Data

Data sources may simply be to old to be useful or relevant to current GIS projects. Past collection standards may be unknown, non-existent, or not currently acceptable. For instance, John Wesley Powell's nineteenth century survey data of the Grand Canyon lacks the precision of data that can be developed and used today. Additionally, much of the information base may have subsequently changed through erosion, deposition, and other geomorphic processes. Despite the power of GIS, reliance on old data may unknowingly skew, bias, or negate results.

Areal Cover

Data on a give area may be completely lacking, or only partial levels of information may be available for use in a GIS project. For example, vegetation or soils maps may be incomplete at borders and transition zones and fail to accurately portray reality. Another example is the lack of remote sensing data in certain parts of the world due to almost continuous cloud cover. Uniform, accurate coverage may not be available, and the user must decide what level of generalization is necessary, or whether further collection of data is required.

Map Scale

The ability to show detail in a map is determined by its scale. A map with a scale of 1:1000 can illustrate much finer points of data than a smaller scale map of 1:250000. Scale restricts type, quantity, and quality of data. One must match the appropriate scale to the level of detail required in the project. Enlarging a small scale map does not increase its level of accuracy or detail.

Density of Observations

The number of observations within an area is a guide to data reliability and should be known by the map user. An insufficient number of observations may not provide the level of resolution required to adequately perform spatial analysis and determine the patterns GIS projects seek to resolve or define. A case in point, if the contour line interval on a map is 40 feet, resolution below this level is not accurately possible. Lines on a map are a generalization based on the interval of recorded data, thus the closer the sampling interval, the more accurate the portrayed data.

Relevance

Quite often the desired data regarding a site or area may not exist, and "surrogate" data may have to be used instead. A valid relationship must exist between the surrogate and the phenomenon it is used to study but, even then, error may creep in because the phenomenon is not being measured directly. A local example of the use of surrogate data are habitat studies of the golden-cheeked warblers in the Hill Country. It is very costly (and disturbing to the birds) to inventory these habitats through direct field observation. But the warblers prefer to live in stands of old growth cedar *Juniperus ashei*. These stands can be identified from aerial photographs. The density of *Juniperus ashei* can be used as surrogate measure of the density of warbler habitat. But, of course, some areas of cedar may uninhabited or inhibited to a very high density. These areas will be missed when aerial photographs are used to tabulate habitats.

Another example of surrogate data are electronic signals from remote sensing that are use to estimate vegetation cover, soil types, erosion susceptibility, and many other characteristics. The data is being obtained by an indirect method. Sensors on the satellite do not "see" trees, but only certain digital signatures typical of trees and vegetation. Sometimes these signatures are recorded by satellites even when trees and vegetation are not present (false positives) or not recorded when trees and vegetation are present (false negatives). Due to cost of gathering on site information, surrogate data is often substituted, and the user must understand variations may occur, and although assumptions may be valid, they may not necessarily be accurate.

Format

Methods of formatting digital information for transmission, storage, and processing may introduce error in the data. Conversion of scale, projection, changing from raster to vector format, and resolution size of pixels are examples of possible areas for format error. Expediency and cost often require data reformation to the "lowest common denominator" for transmission and use by multiple GIS. Multiple conversions from one format to another may create a ratchet effect similar to making copies of copies on a photo copy machine. Additionally, international standards for cartographic data transmission, storage and retrieval are not fully implemented.

Accessibility

Accessibility to data is not equal. What is open and readily available in one country may be restricted, classified, or unobtainable in another. Prior to the break-up of the former Soviet Union, a common highway map that is taken for granted in this country was considered classified information and unobtainable to most people. Military restrictions, inter-agency rivalry, privacy laws, and economic factors may restrict data availability or the level of accuracy in the data.

Cost

Extensive and reliable data is often quite expensive to obtain or convert. Initiating new collection of data may be too expensive for the benefits gained in a particular GIS project and project managers must balance their desire for accuracy the cost of the information. True accuracy is expensive and may be unaffordable.

Errors Resulting from Natural Variation or from Original Measurements

Although these error sources may not be as obvious, careful checking will reveal their influence on the project data.

Positional Accuracy

Positional accuracy is a measurement of the variance of map features and the true position of the attribute. It is dependent on the type of data being used or observed. Mapmakers can accurately place well-defined objects and features such as roads, buildings, boundary lines, and discrete topographical units on maps and in digital systems, whereas less discrete boundaries such as vegetation or soil type may reflect the estimates of the cartographer. Climate, biomes, relief, soil type, drainage, and other features lack sharp boundaries in nature and are subject to interpretation. Faulty or biased field work, map digitizing errors and conversion, and scanning errors can all result in inaccurate maps for GIS projects.

Accuracy of Content

Maps must be correct and free from bias. Qualitative accuracy refers to the correct labeling and presence of specific features. For example, a pine forest may be incorrectly labeled as a spruce forest, thereby introducing error that may not be known or noticeable to the map or data user. Certain features may be omitted from the map or spatial database through oversight, or by design.

Other errors in quantitative accuracy may occur from faulty instrument calibration used to measure specific features such as altitude, soil or water pH, or atmospheric gases. Mistakes made in the field or laboratory may be undetectable in the GIS project unless the user has conflicting or corroborating information available.

Sources of Variation in Data

Variations in data may be due to measurement error introduced by faulty observation, biased observers, or by miscalibrated or inappropriate equipment. For example, one can not expect sub-meter accuracy with a hand-held, non-differential GPS receiver. Likewise, an incorrectly calibrated dissolved oxygen meter would produce incorrect values of oxygen concentration in a stream.

There may also be a natural variation in data being collected, a variation that may not be detected during collection. As an example, salinity in Texas bays and estuaries varies during the year and is dependent upon freshwater influx and evaporation. If one was not aware of this natural variation, incorrect assumptions and decisions could be made, and significant error introduced into the GIS project. In any case, if the errors do not lead to unexpected results, their detection may be extremely difficult.

Errors Arising through Processing

Processing errors are the most difficult to detect by GIS users and must be specifically looked for and require knowledge of the information and the systems used to process it. These are subtle errors that occur in several ways, and are therefore potentially more insidious, particularly because they can occur in multiple sets of data being manipulated in a GIS project.

Numerical Errors

Different computers may not have the same capability to perform complex mathematical operations and may produce significantly different results for the same problem. Burrough (1990) cites an example in number squaring that produced 1200% difference. Computer processing errors occur in rounding off operations and are subject to the inherent limits of number manipulation by the processor. Another source of error may from faulty processors, such as the recent mathematical problem identified in Intel's Pentium (tm) chip. In certain calculations, the chip would yield the wrong answer.

A major challenge is the accurate conversion of existing to maps to digital form. Because computers must manipulate data in a digital format, numerical errors in processing can lead to inaccurate results. In any case, numerical processing errors are extremely difficult to detect, and perhaps assume a sophistication not present in most GIS workers or project managers.

Errors in Topological Analysis

Logic errors may cause incorrect manipulation of data and topological analyses. One must recognize that data is not uniform and is subject to variation. Overlaying multiple layers of maps can result in problems such as Slivers, Overshoots, and Dangles. Variation in accuracy between different map layers may be obscured during processing leading to the creation of "virtual data which may be difficult to detect from real data" .

Classification and Generalization Problems

For the human mind to comprehend vast amounts of data, it must be classified, and in some cases generalized, to be understandable. According to Burrough, about seven divisions of data is ideal and may be retained in human short-term memory. Defining class intervals is another problem area. For instance, defining a cause of death in males between 18-25 years old would probably be significantly different in a class interval of 18-40 years old. Data is most accurately displayed and manipulated in small multiples. Defining a reasonable multiple and asking the question "compared to what" is critical. Classification and generalization of attributes used in GIS are subject to interpolation error and may introduce irregularities in the data that is hard to detect.

Digitizing and Geocoding Errors

Processing errors occur during other phases of data manipulation such as digitizing and geocoding, overlay and boundary intersections, and errors from rasterizing a vector map. Physiological errors of the operator by involuntary muscle contractions may result in spikes, switchbacks, polygonal knots, and loops. Errors associated with damaged source maps, operator error while digitizing, and bias can be checked by comparing original maps with digitized versions. Other errors are more elusive.

The Problems of Propagation and Cascading

This discussion focused to this point on errors that may be present in *single* sets of data. GIS usually depend on comparisons of *many* sets of data. This schematic diagram shows how a variety of discrete datasets may have to be combined and compared to solve a resource analysis problem. It is unlikely that the information contained in each layer is of equal accuracy and precision. Errors may also have been made compiling the information. If this is the case, the solution to the GIS problem may itself be inaccurate, imprecise, or erroneous.

The point is that inaccuracy, imprecision, and error may be compounded in GIS that employs many data sources. There are two ways in which this compounded may occu.

Propagation

Propagation occurs when one error leads to another. For example, if a map registration point has been mis-digitized in one coverage and is then used to register the second coverage, the second coverage will propagate the first mistake. In this way, a single error may lead to others and spread until it corrupts data throughout the entire GIS project. To avoid this problem, use the largest scale map to register your points.

Often propagation occurs in an additive fashion, as when maps of different accuracy are collated.

Cascading

Cascading means that erroneous, imprecise, and inaccurate information will skew a GIS solution when information is combined selectively into new layers and coverages. In a sense, cascading occurs when errors are allowed to propagate unchecked from layer to layer repeatedly.

The effects of cascading can be very difficult to predict. They may be additive or multiplicative and can vary depending on how information is combined, that is from situation to situation. Because cascading can have such unpredictable effects, it is important to test for its influence on a given GIS solution. This is done by calibrating a GIS database using techniques such as sensitivity analysis. Sensitivity analysis allows the users to gauge how and how much errors will affect solutions.

It is also important to realize that propagation and cascading may affect horizontal, vertical, attribute, conceptual, and logical accuracy and precision.

Beware of False Precision and False Accuracy.

GIS users are not always aware of the difficult problems caused by error, inaccuracy, and imprecision. They often fall prey to False Precision and False Accuracy, that is they report their findings to a level of precision or accuracy that is impossible to achieve with their source materials. If locations on a GIS coverage are only measured within a hundred feet of their true position, it makes no sense to report predicted locations in a solution to a tenth of a foot. That is, just because computers can store numeric figures down many decimal places does not mean that all those decimal places are "significant." It is important for GIS solutions to be reported honestly, and only to the level of accuracy and precision they can support.

This means in practice that GIS solutions are often best reported as ranges or ranking, or presented within statistical confidence intervals.

The Dangers of Undocumented Data

It is easy to understand the dangers of using undocumented data in a GIS project. Unless the user has a clear idea of the accuracy and precision of a dataset, mixing this data into a GIS can be very risky. Data that you have prepared carefully may be disrupted by mistakes someone else made. This brings up three important issues.

Ask or look for metadata or data quality reports when you borrow or purchase data.

Many major governmental and commercial data producers work to well-established standards of accuracy and precision that are available publicly in printed or digital form. These documents will tell you exactly how maps and datasets were compiled and such reports should be studied carefully. Data quality reports are usually provided with datasets obtained from local and state government agencies or from private suppliers. Prepare a data quality report for datasets you create.

Your data will not be valuable to others unless you too prepare a data quality report. Even if you do not plan to share your data with others, you should prepare a report – just in case you use the dataset again in the future. If you do not document the dataset when you create it, you may end up wasting time later having to check it a second time. Use the data quality reports found above as models for documenting your dataset.

Metadata

Metadata is data about data. It is a summary document providing content, quality, type, creation, and spatial information about a dataset. Let's take an example. You visit a car dealership to purchase a car. On the window of each car is a sticker giving you very specific information about the vehicle including manufacturer, make, model, size of engine, transmission type, miles per gallon, accessories, etc. This is metadata about the characteristics of a specific vehicle. It is the information you use to make an informed decision when comparing and purchasing a vehicle. Without this information, you know nothing about the vehicle and your decision to purchase becomes confusing at best. This is also true for GIS data. If you don't know what it represents, what it covers, who made it or what quality it is, then only the originator of the data would be able to find and use it. If you do find it and use it, it may be totally inappropriate for your project and give you erroneous results.

Metadata can make clear to users the quality of a dataset or service and what it contains. Based on the metadata, you can then decide whether a dataset or service is useful or not, or whether you need to collect additional data. If the data has a metadata file, the knowledge about the data and services does not disappear if the originator of the data is no longer associated with the data.

It is not necessary for metadata to always give access to the dataset or service; however, it must always indicate where the dataset or service can be obtained.

Digitizing Errors in GIS

Digitizing in GIS is the process of converting geographic data either from a hardcopy or a scanned image into vector data by tracing the features. During the digitzing process, features from the traced map or image are captured as coordinates in either point, line, or polygon format.

Types of Digitizing in GIS

There are several types of digitizing methods. Manual digitizing involves tracing geographic features from an external digitizing tablet using a puck (a type of mouse specialized for tracing and capturing geographic features from the tablet). Heads up digitizing (also referred to as on-screen digitizing) is the method of tracing geographic features from another dataset (usually an aerial, satellite image, or scanned image of a map) directly on the computer screen. Automated digitizing involves using image processing software that contains pattern recognition technology to generated vectors.

Types of Digitizing Errors in GIS

Since most common methods of digitizing involve the interpretation of geographic features via the human hand, there are several types of errors that can occur during the course of capturing the data. The type of error that occurs when the feature is not captured properly is called a positional error, as opposed to attribute errors where information about the feature capture is inaccurate or false.

During the digitizing process, vectors are connected to other lines by a node, which marks the point of intersection. Vertices are defining points along the shape of an unbroken line. All lines have a starting point known as a starting node and an ending node. If the line is not a straight line, then any bends and curves on that line are defined by vertices (vertex for a singular bend). Any intersection of two lines is denoted by node at the point of the intersection.

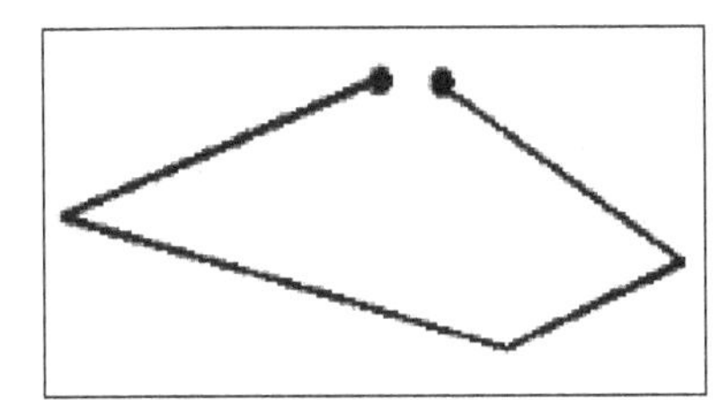

An open polygon caused by the endpoints not snapping together.

Switchbacks, Knots and Loops

These types of errors are introduced when the digitizer has an unsteady hand and moves the cursor or puck in such a way that the line being digitized ends up with extra vertices and/or nodes. In the case of switchbacks, extra vertices are introduced and the line ends up with a bend in it. With knots and loops, the line folds back onto itself, creating small polygon like geometry known as weird polygons.

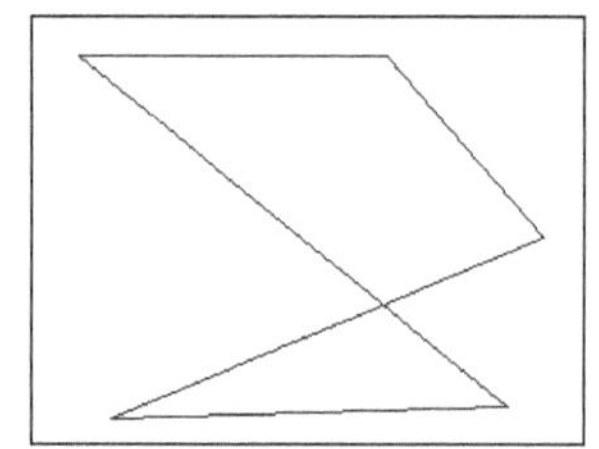

Example of a weird polygon where the line folds back on itself.

Overshoots and Undershoots

Similar to dangles, overshoots and undershoots happen when the line digitized doesn't connect properly with the neighboring line it should intersect with. During digitization a snap tolerance is set by the digitizer. The snap tolerance or snap distance is the measurement of the diameter extending from the point of the cursor. Any nodes of neighboring lines that fall within the circle of the snap tolerance will result in the end points of the line being digitized automatically snapping to the nearest node. Undershoots and overshoots occur when the snap distance is either not set or is set too low for the scale being digitized. Conversely, if the snap distance is set too high and the line endpoint snaps to the wrong node. In a few cases, undershoots and overshoots are not actually errors. One instance would be the presence of cul-de-sacs (i.e. dead ends) within a road GIS database.

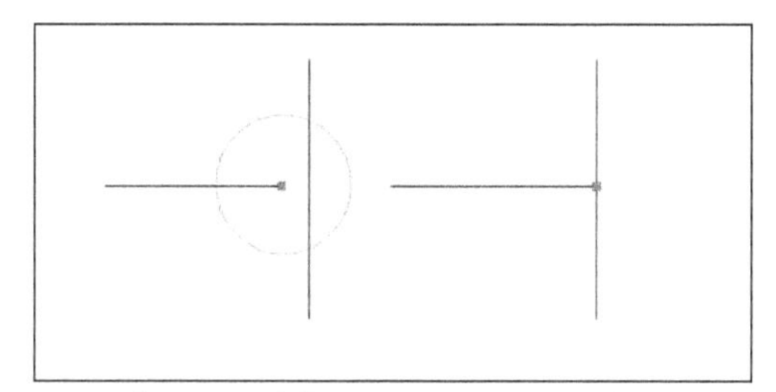

The circle represents: The area of the snap tolerance. The line being digitized will automatically snap to the nearest nodes within the snap tolerance area.

Slivers

Slivers are gaps in a digitized polygon layer where the adjoining polygons have gaps between them. Again, setting the proper parameters for snap tolerance is critical for ensuring that the edges of adjoining polygons snap together to eliminate those gaps. Where the two adjacent polygons overlap in error, the area where the two polygons overlap is called a sliver.

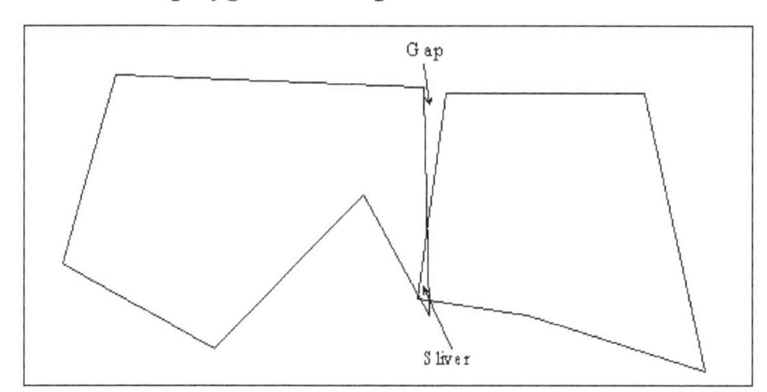

Gap and sliver errors in digitized polygons.

References

- Data-models-for-gis, text-essentials-of-geographic-information-systems: github.io, Retrieved 12 July, 2019
- Flanagin, a. J.; metzger, m. J. (2008). "the credibility of volunteered geographic information". Geojournal. 72 (3–4): 137–148. Doi:10.1007/s10708-008-9188-y
- Spatial-data-model, introduction-gis, lessons, gis250: washington.edu, Retrieved 1 August, 2019
- Elwood, s. (2008). "volunteered geographic information: future research directions motivated by critical, participatory, and feminist gis". Geojournal. 72 (3&4): 173–183. Citeseerx 10.1.1.464.751. Doi:10.1007/s10708-008-9186-0
- What-is-geodata-geospatial-data: gisgeography.com, Retrieved 5 March, 2019
- What-is-gis-mapping-a-beginners-guide-for-site-selectors-and-strategists: buxtonco.com, Retrieved 30 April, 2019
- Goodchild, m.f. (2007). "citizens as sensors: the world of volunteered geography". Geojournal. 69 (4): 211–221. Citeseerx 10.1.1.525.2435. Doi:10.1007/s10708-007-9111-y
- Gis-maps-environmental-monitoring, geographic-information-systems, services: satimagingcorp.com, Retrieved 16 January, 2019
- what-is-mapping-charting-and-visualization, production-mapping, extensions, latest, arcmap: arcgis.com, Retrieved 29 March, 2019

3

Techniques, Tools and Software in Geographic Information System

Various kinds of techniques, tools and software are used in geographic information system. Some of these are remote sensing, hydrological modelling, cartographical modelling, geoprocessing tools, spatial join tool, etc. This chapter discusses in detail these techniques, tools and software related to geographic information system.

GIS DATA CAPTURE

GIS data capture is a technique in which the information on various map attributes, facilities, assets, and organizational data are digitized and organized on a target GIS system on appropriate layers. GIS Data capturing using following features:

Primary GIS Data Capture Techniques which uses Remote Sensing and Surveying Techniques

- Raster data capture: Capturing of attributes, etc without physical contact. This is usually done with the help of satellite imaging techniques, Ariel photography. The advantage of such a GIS data capture is that there is a consistency in the data generated, and the whole process can be regularly and systematically manifolded to get accuracy of the data at a very cost effective manner.
- Vector data capture: This includes capturing of data-sets through physical surveying techniques such as DPGS (Differential Global Positioning System) and Electronic Total Station (ETS). Although this technique is the most effective process to have the accurate data on the target GIS system, it is more time consuming and expensive.

Secondary Data Capture Techniques

- Scanning the raster data: High resolution scanners give very accurate raster images from the hard copies, which can be geo-referenced, and digitized to get the vector output.
- Manual digitizing: Digitization is done directly over the raster by the use of a digitizing tablet, which is a manual pointing device that creates an identical vector map on the computer screen, defining the vertices, points, line data, etc.

- Heads-up digitizing: This is similar to the manual digitization, but the raster scanned data is imported and laid below the vector data to be traced on the computer screen itself.
- Automatic raster to vector conversion: With the advancements of the technology, special software using intelligent algorithms have been developed to recognize the patterns of the points, lines and polygon features and capture them automatically to generate vector GIS data.
- Photogrammetry: Digital stereo-plotters are used to capture the vector data from the Ariel photographs, pictures and images. This is comparatively the most effective method of accurate GIS data capture, but is one of the most costliest methods too.

GIS Data Capture is used in Varied Fields such as for:

- Thematic Maps: Analyzing practical regional/cultural issues, transportation facilitation, hydro graphic mapping, vegetation and other types of related features.
- Electrical power networks are captured using special software for GIS data capture.
- Navigation data are captured for easy navigation purposes.
- Land records and survey data are captured for property, land, water and holding tax, etc. The spatial features are extracted from Ariel imagery using photo-grammetry methods.
- Utility infrastructure GIS data capture for water lines, road network, pavements, sewerage network, and other related features.
- Environmental and geological GIS Data capture is done from geological maps, weather maps, mining and mineral exploration maps, etc.

GEOPORTAL

A geoportal is a gateway to Web-based geospatial resources, enabling users to discover, view and access geospatial information and services made available by their providing organizations. Likewise, data providers can use the geoportal to make their geospatial resources discoverable, viewable, and accessible to others.

What is the Geoportal Server?

The Geoportal Server is a suite of software modules that together allow an organization to build a custom geoportal that meet its style, resource needs, and use objectives. The Geoportal Server provides seamless communication with data services that use a wide range of communication protocols, and also supports searching, publishing, and managing standards-based resources.

The suite of software modules includes the following:

- A customizable geoportal web application for publishing, administering, and searching resources.

- A live data previewer map interface for viewing live resources.
- Integration with content management systems to organize resources to support focused user communities.
- Data extraction service customization for downloading data for a resource, with the ability to specify an extent, projection and download format.
- Search results exposed through the REST API so resources can be easily shared among applications and users.
- CSW Client, a freely downloadable extension for ArcMap and ArcGIS Explorer to enable searching geoportal catalogs from within those environments.
- Publish Client, a freely downloadable extension for ArcCatalog that enables geoportal publisher users to publish resources from an ArcCatalog directory to the geoportal.
- WMC Client, a freely downloadable extension for opening saved web map context files in ArcMap.
- Widgets for searching geoportals from an HTML page, a Flex-based viewer, or a Silverlight-based viewer.

The basic components of a Geoportal Server implementation are shown below.

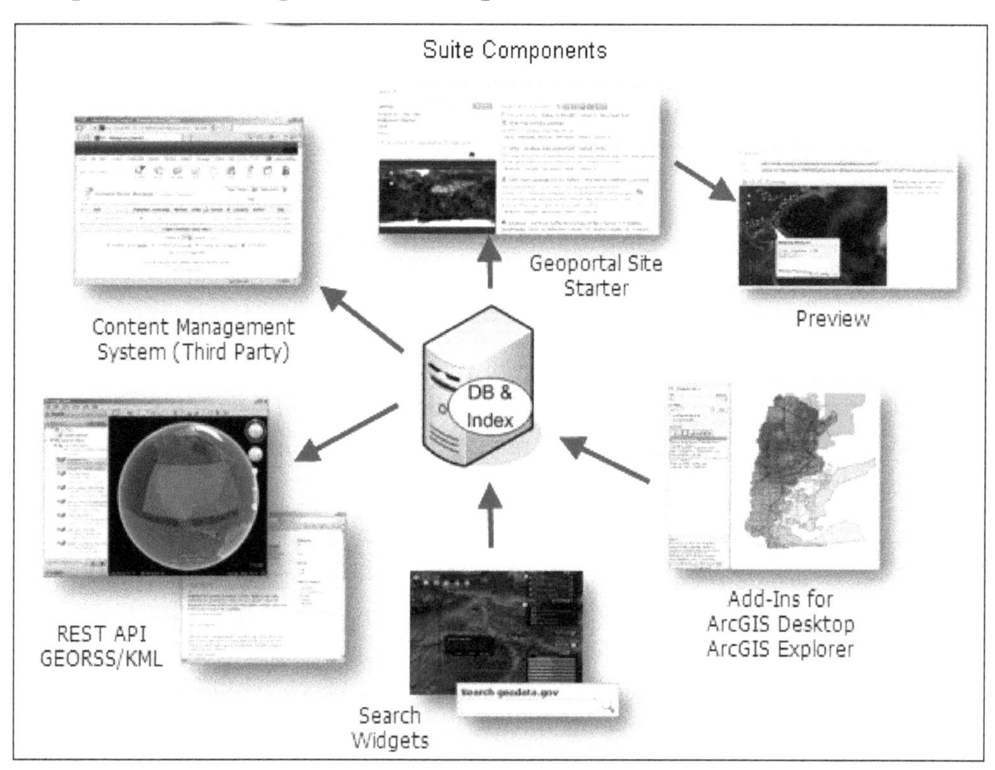

What is a Resource?

A resource - in geoportal terms - is information, data, or a repository that hosts information or data.

There are a number of things that can be done with a resource using the Geoportal Server. Users can discover resources through the geoportal and read metadata about the resource. Users can share the resource with other users by sending its REST API URL through an email, Twitter feed, social network posting, or content management system link. Providers of the resource can publish it to the geoportal so users can discover the resource and more like it.

GEOPROCESSING TOOLS

The Buffer Tool

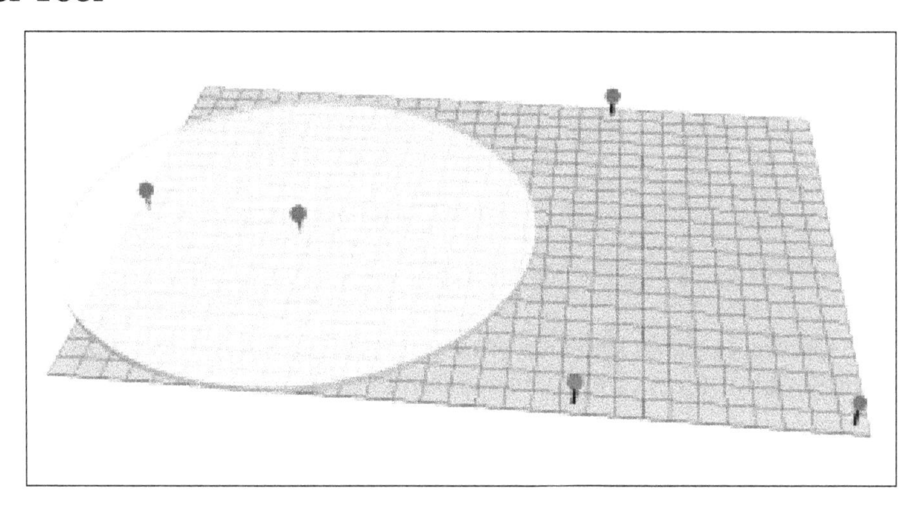

Buffers are proximity functions. When you use this geoprocessing tool, it creates a polygon at a set distance surrounding the features.

For example, a buffer is a polygon or collection of cells that are within a specified proximity of a set of features.

Buffers can have fixed and variable distances. In addition, they can be set to geodesic which accounts for the curvature of the Earth.

Buffer Example: Chernobyl Exclusion Zone

Chernobyl is the worst nuclear disaster in human history. In a short period of time, it released hundreds of times more radiation than Hiroshima.

Also, it is one of only two classified as a level 7 event (the maximum classification). Surrounding vegetation absorbed radioactive isotopes and died within a week of the blast.

As a result of the deadly toxins released in the atmosphere, safety crew declared a 2600 square kilometer buffer around the nuclear power plant. To this day, this buffer zone is still in effect and it's called the Chernobyl exclusion zone.

30 years later, the trees remain reddish-brown. There is an estimated 9000 to 93,000 deaths across Europe. And the exclusion zone is still in effect.

The point of the story is that if ArcGIS was around at the time, they could've ran the buffer geoprocessing tool. Since the blast, satellites like SPOT have been monitoring the Chernobyl exclusion zone because of its restrictions.

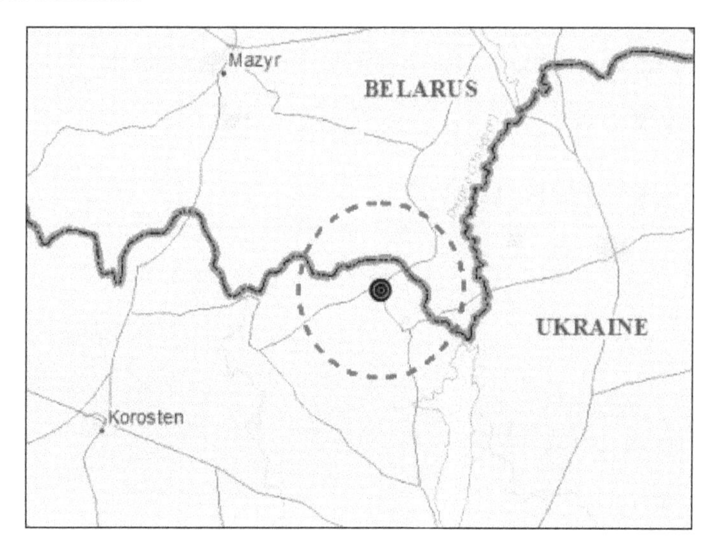

The Clip Tool

The clip tool is an overlay function that cuts out an input layer with the extent of a defined feature boundary. The result of this tool is a new clipped output layer.

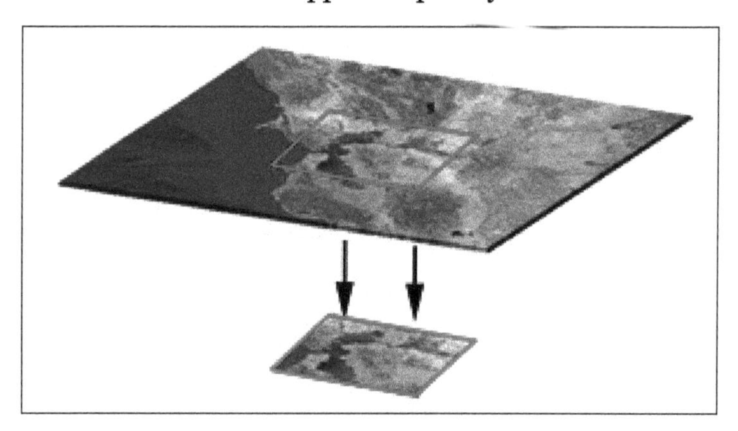

If you can picture a cookie cutter, this is like using the clip tool. And carving out vectors and rasters is one of the most common operations in GIS.

In order to clip data, you need points, lines or polygons as input and a polygon as the clipping extent. The preserved data is the result of a clip.

Clip Example: Carving out Florida

Florida's nickname is the sunshine state. You can even find the Sunshine State on their license plate. But how much sunshine does Florida really receive?

It turns out that global horizontal irradiance (GHI) is a good measure of incoming solar radiation. So if you wanted to install a solar panel, GHI is the recommended data set.

If you clip GHI to the Florida state boundary, you can really find how much sunshine Florida really gets. When we clip GHI, we can add it to a map of even summarize the average GHI values.

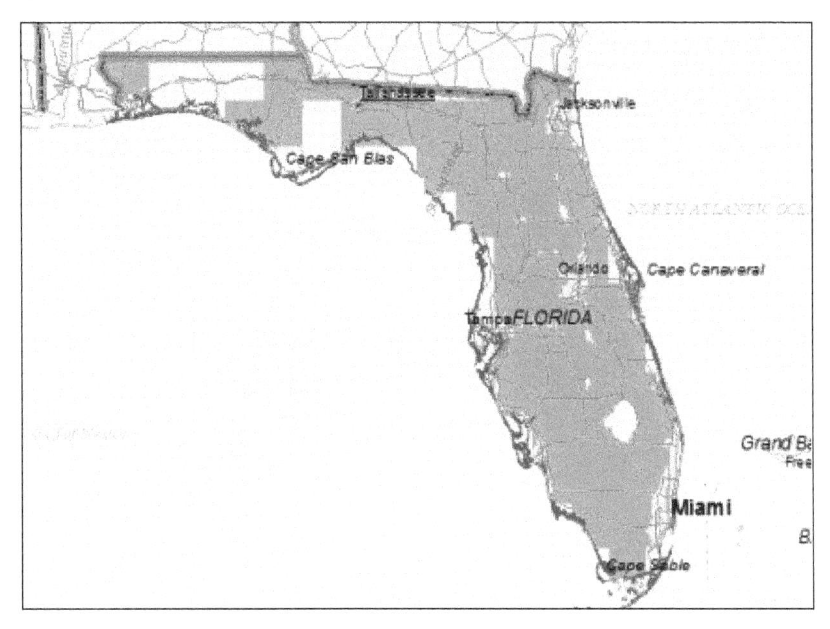

The Merge Tool

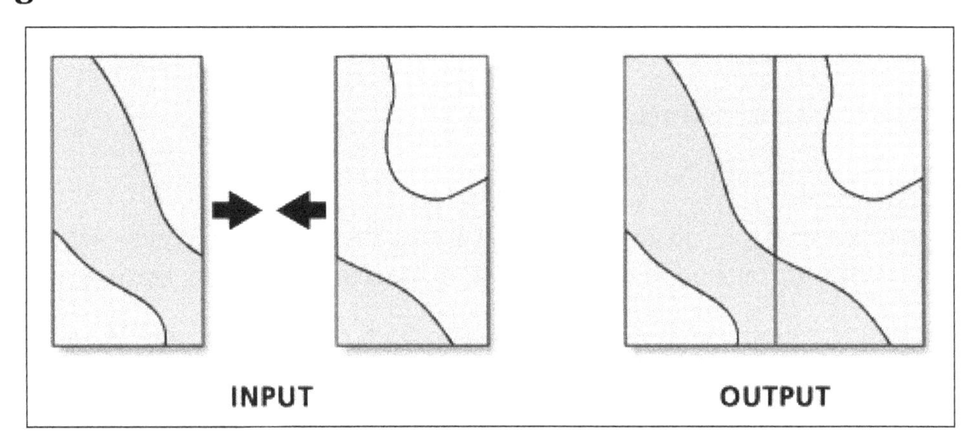

The merge geoprocessing tool combines data sets that are the same data type (points, lines or polygons). When you run the merge tool, the resulting data will be merged into one.

Similar to clip, we use the merge tool on a regular basis. For merging, data sets have to be the same type. For example, you can't merge points and polygons into one data set.

Merge Example: The Supermarket Merger

For example, if two grocery store giants like Ahold NV and Delhaize Group want to combine their 6,500 stores, we can use the merge tool.

In this case, we have two existing data sets from both companies. The merger between the two grocery stores into one company – Ahold Delhaize – means all grocery stores will be combined into a single data set. When you combine grocery stores (points) from both companies, they all end up in a final data set.

The Dissolve Tool

The Dissolve Tool unifies boundaries based on common attribute values. In other words, dissolve merges neighboring boundaries if the neighbors have the same attributes.

For example, if you want to remove the borders of countries to form a continent, the dissolve tool is the tool to use. But you would need an attribute for each country and the continent it belongs to.

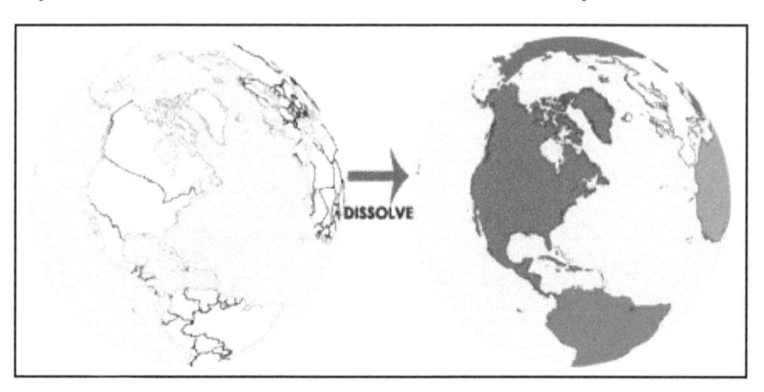

Dissolve Example: Unifying Countries

What do Germany, Yemen, Tanzania and Vietnam all share in common? They are all examples of two countries dissolving their borders and unifying to form one. Country unification is a rare event. But dissolving boundaries in GIS is not.

West Germany + East Germany = Germany

North Vietnam + South Vietnam = Vietnam

The dissolve geoprocessing tool erases borders and unifies them into one. When each country has its continent name in the attribute table, you can run the dissolve tool to unify borders into continents.

Over 25 years ago, the Berlin Wall was wiped away which divided East and West. The East and West dissolved its walls into a single country.

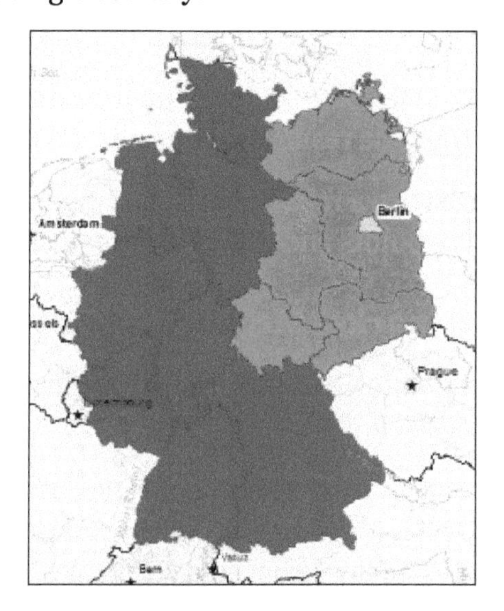

The Intersect Tool

The Intersect Tool is very similar to the clip tool because the extents of input features defines the output. The only exception is that it preserves attributes from all the data sets that overlap each other in the output.

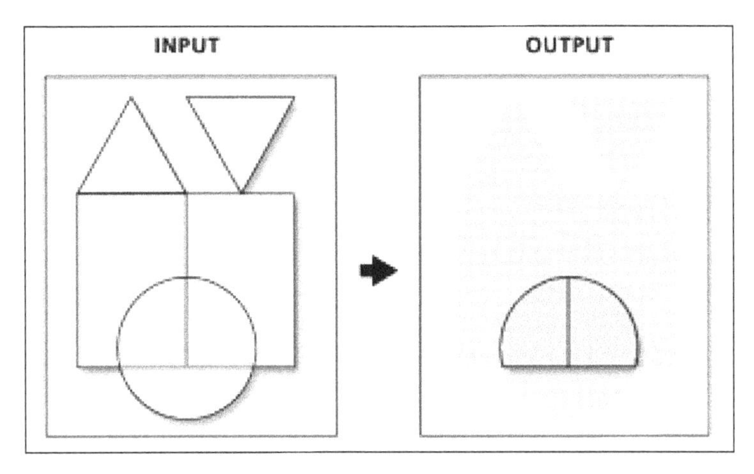

The Intersect Tool performs a geometric overlap. All features that overlap in all layers will be part of the output feature class – attributes preserved.

Add multiple inputs. The tool accepts different data types (points, lines and polygons). When features overlap each other, they will be in the output. The Intersect Tool preserves the attribute values in both input layers.

Run the Intersect Tool on a single feature and you can find overlaps.

Intersect Example: Generating Pivot Tables

The city councilor asked the GIS analyst: "How many apartments, condos and houses do we have in precinct A, B and C? Create a pivot table for me."

Instead of running a clip, it would be helpful to run an intersect. Why? Because we preserve attributes from both input data sets. You need the building type from the dwellings layer. You need the precinct ID from the residential layer.

When you run an intersect with the dwellings and residential layers, the output will have all the points that overlap for each precinct. Most importantly, it will keep the dwelling type AND precinct ID.

Select all the rows. Ctrl-C in ArcGIS. Ctrl-V in Excel. Select all. Insert pivot table.

Run the Tabulate Intersection Tool in ArcGIS.

What's the difference between the clip tool and the intersect tool?

The main difference is the resulting attributes. When you run the clip tool, only the input features attributes will be in the output. When you use the intersect tool, the attributes from all features will be in the output.

The Union Tool

Some say the Union tool should come with a bottle of antacid. The union tool gets a bad reputation because it creates a lot of features. The Union Tool maintains all input features boundaries and attributes in the output feature class.

After running this geoprocessing tool, it does get a bit messy especially when there are more overlaps. But it's really not so bad. The Union tool spatially combines two data layers. It preserves features from both layers at the same extents.

Union Example: Basic Shapes

In this example, we have an overlapping circle and square. The circle is a single record and the square is a single record.

When you run a union on these two features, it produces 3 records – the original circle, the original square and the overlapping portion.

Unions have been especially useful in suitability applications because you can understand where different habitat types overlap.

The Erase (Difference) Tool

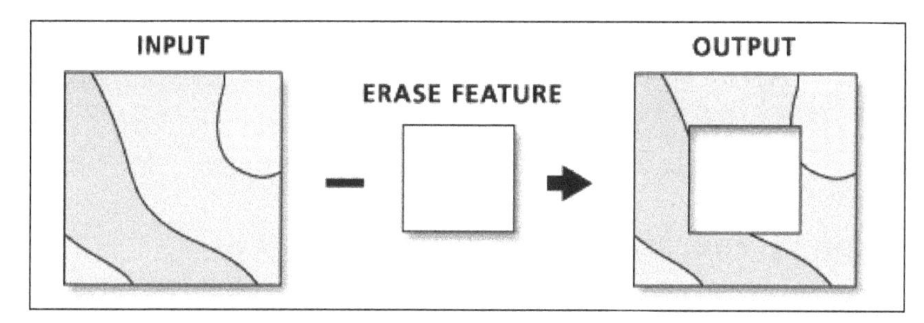

We like the erase tool. It's always been helpful in erasing things. The input layer is what will be erased. The erase feature determines what to erase. Simple as that.

The Erase Tool removes features that overlap the erase features. This geoprocessing tool maintains portions of input features falling outside the erase features extent. The result is a new feature with the erase feature extent removed.

Erase Example: Erasing Forest Burns

Humans start 90% of wildfires. Lightning strikes the Earth 100,000 times a day. 10 to 20% of these lightning strikes can cause a forest fires.

No matter how you slice it, forest fires are happening all times somewhere on Earth. Ecologists need to understand how much suitable habitat exists on the landscape.

When a forest fire tears through a forest, you can run the erase tool because these forest stands no longer exist. They are no longer suitable habitat for certain species. Erase those areas with the wildfire polygons and BOOM, you have an updated habitat extent.

REMOTE SENSING IN GIS

Remote sensing is the science of obtaining information without physically being there. For example, the 3 most common remote sensing methods is by airplane, satellite and drone.

First, if we are going to solve some of the major challenges of our time, remote sensing data is fundamentally important to monitor the Earth as an entire system. Literally, there are thousands of uses for remote sensing.

For example, the Arctic is an unforgiving destination to travel to. Because of the obvious safety risks of field activity, scientists leverage remote sensing for sea ice monitoring, ship tracking and even national defense.

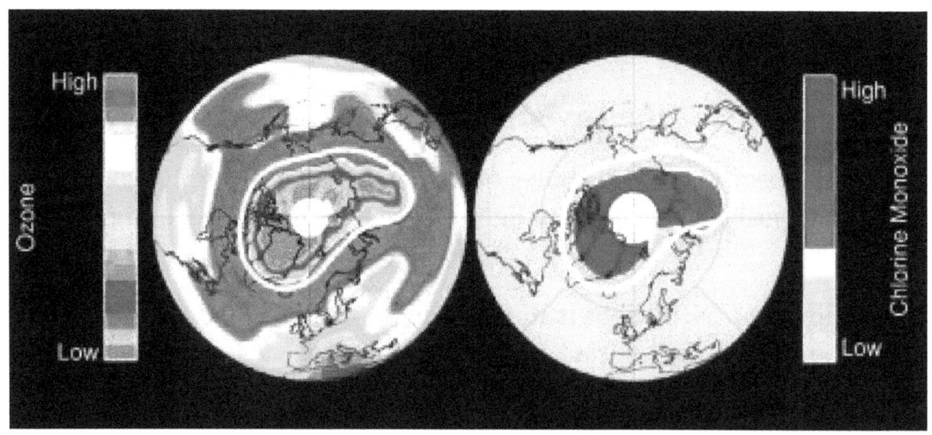

NASA's Aura spacecraft monitors ozone in Earth's stratosphere. While purple delineates low ozone amounts, grey colors display large amounts of chlorine. And dark blue colors describe monoxide.

Already, we see growth in the usage of Light Detection and Ranging (LiDAR) which is simply just a way to accurately measure how far things are. Often, airplanes and drones capture LiDAR depending on the size and area of land. Using digital elevation models from LiDAR, we can better predict risk of flooding, archaeological sites and even autonomous vehicles.

LiDAR stands for Light Detection and Ranging. It uses lasers to measure distance. This is useful for measuring heights on the bare ground and its features.

Because remote sensing covers so much ground, it puts a wealth of information into the hands of decision-makers.

How does Remote Sensing Work?

As mentioned before, remote sensing means you are acquiring information from a distance. When you're outside, the sun emits light. And each object reflects a mix of red, green and blue colors into your eyes. It's the same for sensors on board satellites work.

But what's important to know is that there is a whole range of possible wavelengths in the electromagnetic spectrum from short wavelengths (like X-rays) to long wavelengths (like radio waves).

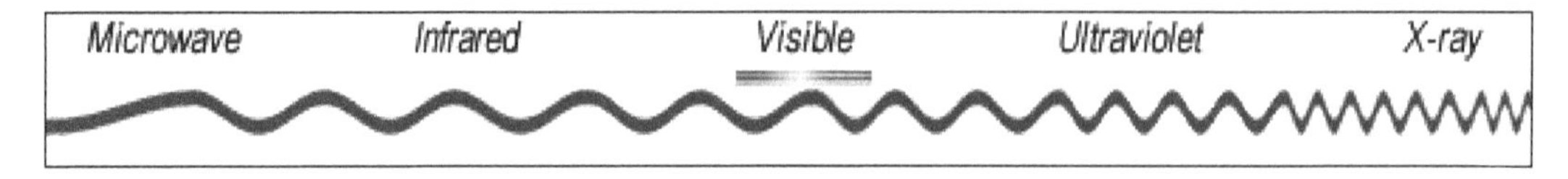

And this is why remote sensing is such a powerful discipline. Because we can see beyond human vision, this fact alone lets us see things we've never seen before. In other words, we can see the invisible.

What does it Mean to See the Invisible?

If you look at the EM spectrum image above, our eyes are sensitive in the visible spectrum (390-700 nm). Really, that's just a small range of light because engineers can design sensors to capture other parts of it.

For example, near-infrared (NIR) light is in the range of 700 to 1400 nm. And we already know that plants reflect more green light because that's how our eyes see it. Using indexes like NDVI, we can classify healthy vegetation for the whole planet.

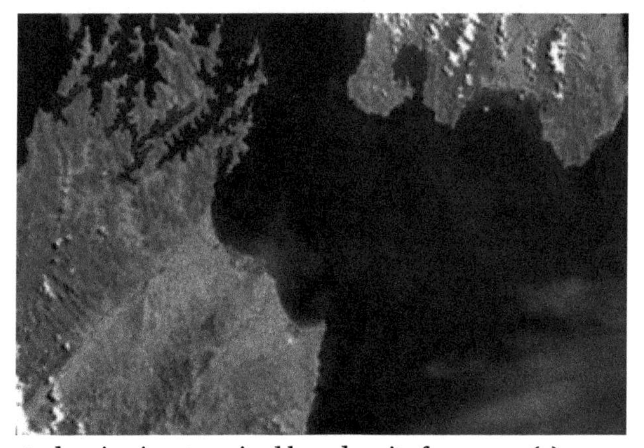

Each spectral region is categorized based on its frequency (v) or wavelength.

Spectral bands are simply groups of wavelengths. Examples of spectral bands are: ultraviolet, visible, near-infrared, mid-infrared, thermal infrared and microwave.

It's not only near-infrared but it turns out that other spectral bands are extremely useful to classify land cover on Earth. We can classify land cover because each object has its own unique spectral signature, depending on its chemical composition.

Because we can use ultraviolet, visible, near-infrared, mid-infrared, thermal infrared, we can classify different features on Earth. Ultimately, land cover is how we understand our changing planet and things like climate change such as this Global Land Survey classification.

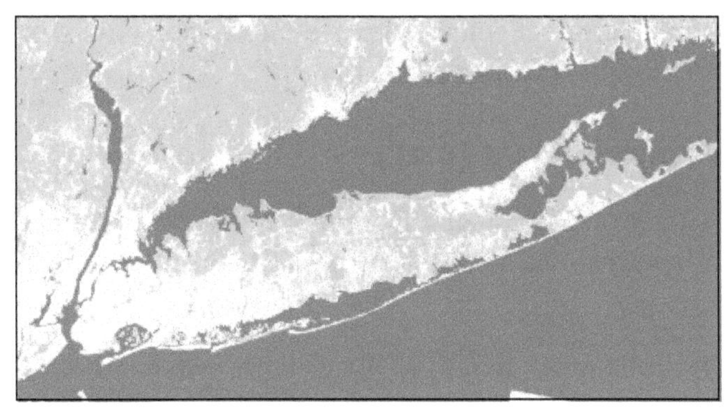

The University of Maryland teamed up with the USGS to create its circa 2010 tree cover, bare ground and persistent surface water.

A spectral signature is the amount of energy reflected in a particular wavelength. Differences in spectral signatures is how we tell objects apart. Spectral signatures are driven by the objects chemical composition.

What are the types of Remote Sensing?

In general, the two types of remote sensing are passive and active remote sening.

Active Sensors have their own source of light or illumination and its sensor measures reflected energy. For example, Radarsat-2 illuminates a target and measures how much energy bounces back to the sensor. Actually, it's similar to the flash of a camera as you can see below.

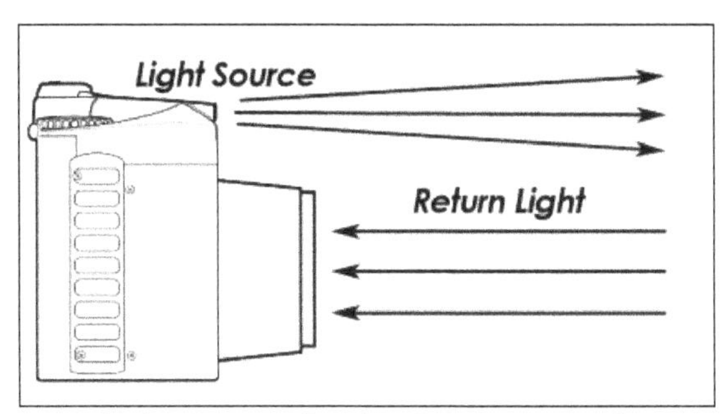

Active Remote Sensing Camera Example.

Alternatively, Passive Sensors measure reflected light emitted from the sun. For example, Landsat-8 measures reflected energy emitted by the sun bouncing off the Earth. But when the flash is turned off, the source of energy comes from somewhere else such as the sun or a lamp.

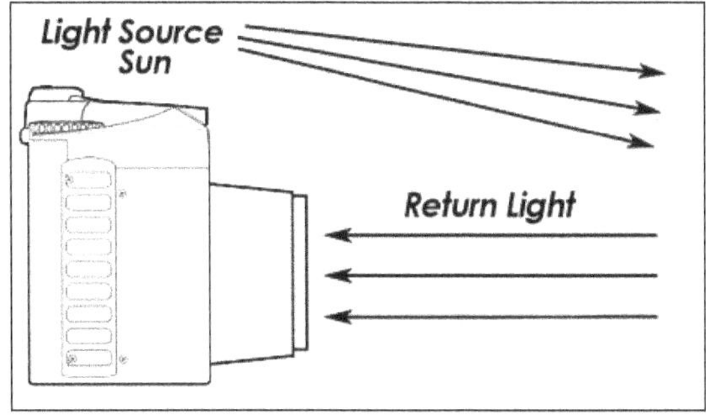

Passive remote sensing camera example.

Passive remote sensing measures reflected energy emitted from the sun, active remote sensing illuminates its target and measures its backscatter.

Multispectral vs. Hyperspectral Imagery

Passive remote sensing can be divided even further into multispectral and hyperspectral. The main difference between multispectral and hyperspectral is the number of bands and how narrow the bands are.

Multispectral imagery generally refers to 3 to 10 bands with each band acquired using a radiometer.

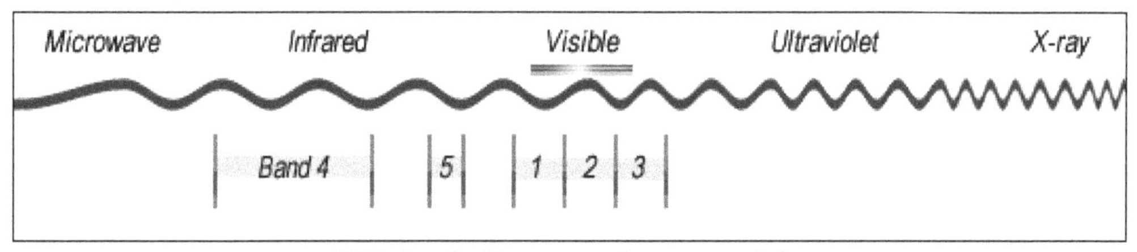

Wide bands (Image not drawn to scale).

Hyperspectral imagery consists of much narrower bands (10-20 nm). A hyperspectral image could have hundreds of thousands of bands using an imaging spectrometer.

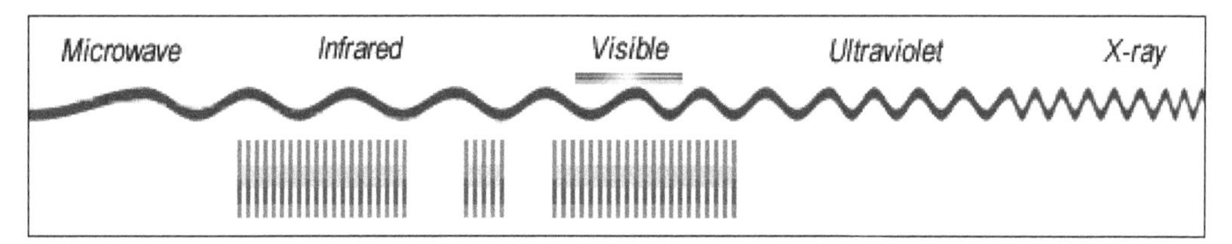

Imagine hundreds of narrow bands (Image not drawn to scale).

When Landsat-8 makes an acquisition in space, it produces 11 separate images for its band designations. In the case of Landsat-8, it consists of

- Coastal aerosol (0.43-0.45 um)
- Blue (0.45-0.51 um)
- Green (0.53-0.59 um)
- Red (0.64-0.67 um)
- Near infrared NIR (0.85-0.88 um)
- Short-wave infrared SWIR 1 (1.57-1.65 um)
- Short-wave infrared SWIR 2 (2.11-2.29 um)
- Panchromatic (0.50-0.68 um)
- Cirrus (1.36-1.38 um)
- Thermal infrared TIRS 1 (10.60-11.19 um)
- Thermal infrared TIRS 2 (11.50-12.51 um)

In terms of hyperspectral satellites, the Hyperion imaging spectrometer (part of the EO-1 satellite) produces 220 spectral bands (0.4-2.5 um). For airborne sensors, NASA's Airborne Visible/Infrared Imaging Spectrometer (AVIRIS) is an example of a hyperspectral sensor delivers 224 contiguous channels with wavelengths from 0.4-2.5 um.

While hyperspectral images has hundreds of narrow bands, multispectral images consist of 3-10 wider bands.

Remote Sensing Image Resolution

Image resolution can be divided into spatial, spectral and temporal resolution.

Spatial Resolution is the detail in pixels of an image. While higher spatial resolution means more detail and smaller pixel size, lower spatial resolution means less detail and larger pixel size.

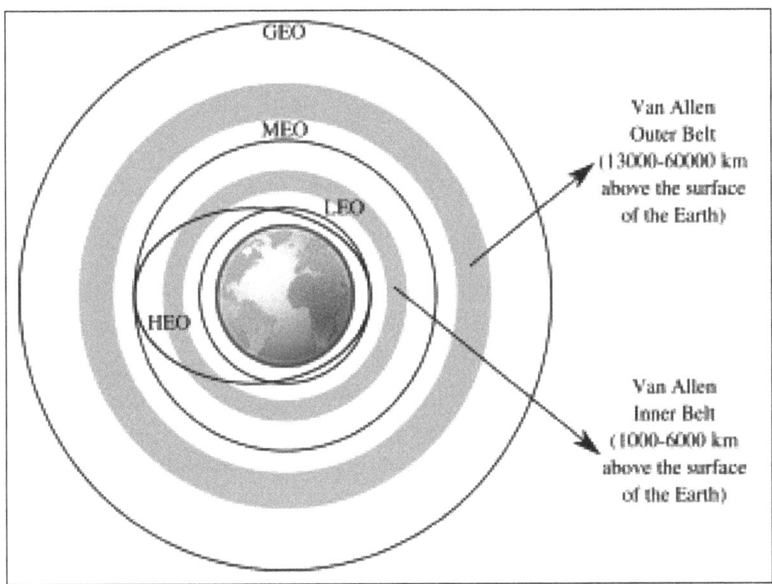

Temporal Resolution is how often revisit times occur, often for satellite data. The height of the satellite above the Earth surface will determine the time it takes for the orbit to take one complete orbit of the Earth. Orbital period increases with satellite height.

Geostationary orbits match the Earth's rate of rotation. Sun Synchronous orbits keeps the angle of sunlight on the surface of the Earth as consistent as possible. Polar orbits passes above or nearly above both poles of Earth.

HYDROLOGICAL MODELLING IN GIS

Hydrology is concerned with study of the motion of the earth's waters through the hydrologic cycle, and the transport of constituents such as sediment and pollutants in the water as it flows. GIS is focused on representing the landscape by means of locationally referenced data describing the character and shape of geographic features. A spatial hydrology model is one which simulates the water flow and transport on a specified region of the earth using GIS data structures. Suppose the boundary of this region is represented by a polygon, such as a river basin boundary or an aquifer boundary. Because both vertical and horizontal water flow can be taking place within the region, hydrologic processes need to be defined over a volume of space rather than an area. Such a volume can be constructed by projecting vertically the lines making up the polygon boundary into the atmosphere and into the earth, and closing the top and bottom of the volume by horizontal planes. Such a volume is called a *control volume* in fluid mechanics and the surface which surrounds it is called the *control surface.*

Building a hydrologic model involves writing equations that relate the rates of change of water properties within the control volume to flow of those properties across the control surface. For example, a simple soil water balance model for a control volume drawn around a block of soil is:

$$S(t+1) = S(t) + P(t) - E(t) - Q(t)$$

in which S(t) represents the amount of soil moisture stored at the beginning of the time interval t, S(t+1) the storage at the end of that interval, and the flow across the control surface during the interval consists of precipitation P(t), evaporation, E(t), and soil moisture surplus, Q(t), which supplies streamflow and groundwater recharge. Solving this equation requires dealing with time series of the four variables: S, P, E, Q, and possibly of other variables related to them.

It is implicit in constructing a spatial hydrology model that the properties of the system will be spatially variable, so a time series for each of the variables just described must be generated for each soil unit in the domain of analysis. Suppose that there are L spatial units, and that analysis on a single unit requires the definition of M variables for each of N time periods. The number of values to be determined is given by the product LMN, a number that can easily explode beyond the capabilities of available computer memory and reasonable computation times. Indeed, it appears that if the product LMN exceeds a limit of approximately 1 million, solution of the model will be computationally difficult. It follows that in constructing a GIS hydrology model the first task is to determine what variables will be calculated on how many spatial units for a defined number of time periods. Construction of complex models must proceed by partitioning the total problem into a series of submodels that interact with a common database. It is the capacity of GIS for rigorously defining this database which makes possible complex models connecting various parts of the hydrologic cycle within a particular region. One begins with a GIS description of the region and then by modeling adds additional detail to the regional description concerning water flow and constituent transport.

Time and Spatial Domains

In considering these questions, a fundamental distinction arises in the treatment of spatial units depending upon whether raster or vector data structures are used. In a raster data structure, a fine mesh of cells is laid over the landscape and all calculations are done for each cell. This approach, based on using digital elevation model cells as the spatial units, is very useful for certain kinds of hydrologic analysis. But the number of DEM cells within typical analysis regions is usually very large, typically 10,000 to 1 million, so the number of time periods that can analyzed is usually relatively small. Indeed, one may eliminate time as a dimension of the problem by working with mean annual values of all the variables, and the subject of raster based mean annual flow and transport models is presented at further length.

If time dynamics are to be considered, it is usually necessary to employ a vector data structure based on related points, lines and polygons. To describe the values on L spatial units of M computed variables in N time periods requires in concept a 3-D data structure but this can be reduced to a set of 2-D data structures in the following manner. The feature attribute table of the GIS coverage defines the geographic properties of the L spatial units and gives each a unique identifying number. This table has geographic attributes as its columns or fields for which the values in each spatial unit are displayed in rows. The values of a particular computed variable, such as soil moisture storage, S, can be defined by means of a related *time table*. The conventional method of constructing such a table is to define a new field for each time interval and keep the rows for the sequence of spatial units, but the advent of object-oriented GIS programming languages, such as the Avenue language in Arcview, has made possible the reverse arrangement, namely the use of time as the index on the rows of the table for which there is a new field for each spatial unit. The item name in this field is the feature identification number of the spatial unit attached to an arbitrary prefix such as SU to mean spatial unit. Thus, in a time table for soil moisture storage, the value of soil moisture storage in time period 154 on spatial unit 30 may be found in row 154 of the table in field SU30.

This arrangement of time vertically and space horizontally makes it easy to see by reading down a column the temporal sequence of values of a variable in a particular spatial unit. The key point in this description is that using object oriented data structures, a value in one table (the feature identification number) can be related to a field in another table instead of to a single value or a set of such values. For hydrologic modeling, the smooth treatment of time variation within GIS is a critical problem that has strongly limited what could be done in the past. It appears that this limitation has now been lifted and the models of spatially distributed and time varying systems can be readily constructed within GIS rather than simply using the GIS as a repository for spatial data feeding an external time varying model.

Ten-step Modeling Procedure

In table is presented a 10-step plan by which preparing a GIS hydrology model can be broken down into component parts. The first five of these steps deal with defining the framework of the model in space, and time, and in preparing the environmental description, which may include representation of the land surface terrain, soils and land cover, subsurface hydrogeology, and hydrologic data such as precipitation, streamflow and constituent concentrations. The second five steps deal with simulating the water balance of spatial units, the flow of water and transport of constituents between units, the effect of water utilization structures such as dams and pumping systems, and finally, with the presentation of the study results.

Table: A ten-step procedure for a GIS hydrology study.

1.	Study design: Objectives and scope of study; spatial and time domain; process models needed, variables to be computed.
2.	Terrain analysis: Deriving a watershed and stream network layout from digital elevation data and mapped streams.
3.	Land surface: Describing soils, land cover, land use, cities, and roads.
4.	Subsurface: Hydrogeologic description of aquifers
5.	Hydrologic data: Locating point gages, attaching time series and their average values, interpolating point climatic data onto grids.
6.	Soil water balance: Partitioning precipitation into evaporation, groundwater recharge and surface runoff; partitioning of chemicals applied to the land surface.
7.	Water flow Movement of water through the landscape in streams and aquifers. Computing streamflow and groundwater flow rates.
8.	Constituent transport: Transport of sediment and contaminants in water as it flows. Computing concentrations and loadings.
9.	Impact of water utilization: Locating reservoirs, water withdrawals and discharges in rivers, and aquifer pumping. Their effects on water flow and constitutent transport.
10.	Presentation of results: Developing visual and tabular presentation of the study results. Use of Internet and CD-ROM to transmit results.

The advent of Statsgo soil data base and the Internet access to the RF1 river reach file, the USGS Hydrologic Unit Code for watershed boundaries and the USGS Land Use and Land Cover data are all a great help in constructing hydrologic models.

Processing Digital Elevation Data

Probably the principal advance in GIS hydrology modeling that has occurred during the past several years has been the widespread availability of digital elevation data via Internet and CD-ROM and advances in the methods of processing them. DEM data are available in the United States at 30m and 3" cell size, suitable for delineating watersheds and stream networks within urban areas or in small river basins, at 15" cell size which is suitable for regional studies of the size of Texas, to 30" cell size which is appropriate for continental scale studies to 5' cell size covering the earth. The 30" DEM's for Africa and other regions being constructed by Sue Jenson and her co-workers at the USGS using the Digital Chart of the World are a significant contribution.

Delineation of watersheds from DEM data has become standardized on the eight-direction pour point model in which each cell is connected to one of its eight neighbor cells (four on the principal axes, four on the diagonals) according to the direction of steepest descent. Given an elevation grid, a grid of flow directions is constructed, and from this is derived a grid of flow accumulation, counting the number of cells upstream of a given cell. Streams are identified as lines of cells whose flow accumulation exceeds a specified number of cells and thus a specified upstream drainage area. Watersheds are identified as the set of all cells draining through a given cell. The "thousand-million" rule is a rough guide in this activity - take the area of the region to be analyzed and divide it by one million to give the appropriate cell size to use; multiply the cell size chosen by one thousand and that is the minimum drainage area of watersheds that should be delineated from this DEM.

Table shows the cell sizes of digital elevation data and their typical range of application. In this table, a typical watershed area contains 5000 DEM cells and the region of application contains 200 of these watersheds.

Table: Digital elevation cell sizes and their scope of application.

Geographic Cell Size	Linear Cell Size	Watershed Area	Region Area	Typical
1"	30m	5	1000	Urban watersheds
3"	90m	40	8000	Rural watersheds
15"	460m	1000	200,000	River basins
30"	930m	4000	900,000	Nations
3'	5.6 km	150,000	30,000,000	Continents
5'	9.3 km	400,000	90,000,000	Global

Standardized Approach to Watershed Delineation

A standardized way of delineating watersheds and stream networks is the following: construct the stream network in the standard way by choosing a drainage area threshold; divide the stream network so created into individual stream links; find the outlet cell at the lower end of each link; delineate the watershed for each of these outlet cells. By changing the threshold drainage area, subwatersheds can be delineated within watersheds in a nicely scaled manner. This algorithm has a critical property - there is one and only one stream for each watershed, which makes modeling of the flow of water from watersheds to the streams within them feasible within GIS. Also, there is a one to one relation between the raster representation of the landscape and the vector features of streams and watersheds derived from it.

Several variations on this standard algorithm are useful. First, the DEM delineated streams are usually close to but do not coincide with mapped streams. Sometimes, critical errors occur when a portion of the upstream part of a basin drains into the wrong downstream river. To overcome these errors, the mapped streams can be converted into a grid and "burned in" to the DEM by artificially raising the elevation of the off-stream cells. This technique requires editing of the stream network to eliminate stray streams and loops that would confuse the delineation process and it results in some distortion of the watershed boundaries in areas where the DEM and mapped streams are not completely consistent, but it has the great advantage that the DEM delineated streams match the mapped streams exactly. After all, it is the stream network which is really the critical item in landscape delineation because it is the stream that carries the water. This "burn in" technique is especially useful in coastal zones with very flat terrain and other locations where drainage is directed through constructed channels. It also helps to ensure that gaging stations and other features are precisely located on the stream. Other variations on the standard delineation technique include the identification of zones of interior drainage and the subdivision of long streams by placing outlet cells along the streams at arbitrarily defined intervals.

Time Averaged Hydrologic Modeling

Mean Annual Flow

The determination of flow at ungauged locations is a common problem in hydrology. A simple approach to this problem is to eliminate time as a dimension by restricting the computation to mean annual flows. The analysis can then be constructed by using the cells of a DEM grid as the computational units. One begins with a mean annual precipitation grid over the landscape, which for the United States has been constructed by Daly et al. and for Africa by Hutchinson et al., both using approximately 3' cells. The precipitation for each DEM cell is determined from the climate grid. The watersheds of each of the stream gauging stations in the region are delineated and the mean annual precipitation, P, for the drainage areas determined. The longest streamflow record in the basin is used as an anchor record, a long period of analysis is chosen, and the mean annual flow per unit of drainage area, Q, is determined for each gage. If some of the gages have incomplete records, the long term estimate of the mean annual flow can be found by: long term flow at a sample gage = long term flow at the anchor gage x (flow at sample gage/flow at anchor gage) where the ratio in parentheses is constructed using the means of the common period of record at the two gages. A graph is plotted of Q versus P, and an equation fitted to the relation:

$$Q = cP$$

where c is a runoff coefficient, which is a nonlinear function of the precipitation. In dry areas, the greater is the precipitation, the greater is the percentage of the precipitation which becomes runoff. By multiplying the mean annual precipitation grid by this runoff coefficient, a mean annual runoff per unit area can be determined for each DEM cell. This quantity can be used as a weight and a weighted flow accumulation performed in the same manner as the regular flow accumulation is done when constructing the watershed boundaries. The weighted flow accumulation of each DEM cell, when multiplied by the cell area, gives the mean annual flow for each cell. Thus a mean annual flow map can be derived with estimates of the flow at every stream location in the landscape. This is a very simplified method of hydrologic analysis but one that is faithful to the gauged data in the region and can be applied to large regions in a consistent manner.

Non-point Source Pollution Assessment

An extension of the mean annual flow estimate just derived can be used to estimate non-point sources of pollution loading to streams. Such sources include pollutants from agricultural areas and from urban runoff from areas such as roads and parking lots. Point pollutant sources are those associated with a particular outlet location, such as the outlet of a wastewater treatment plant. A standard assumption in treating non-point source pollution is to relate the expected concentration of pollutants in runoff to the land use in the drainage basin. By taking a land use map of the area, and using a look-up table to connect land use to expected pollutant concentration, a map of expected concentration, C, can be determined and the expected pollutant concentration in runoff from each DEM cell calculated. By taking the product of the concentration and the flow rate per unit area, a pollutant loading rate per unit area, L, can be calculated for each cell using the relationship:

$$L=CQ$$

Computing the weighted flow accumulation of this loading onto downstream cells and multiplying by the cell area gives the mean annual pollutant loading in each cell. Nice maps of expected pollutant loadings can thus be created. By dividing this mean annual loading by the mean annual flow, a grid of expected pollutant concentration in each cell is derived, which is derived from the weighted average of all the contributions of flow and loading from upstream. Thus is created an expected concentration map, which of course is really meaningful only in the stream cells.

Pollutant concentration is normally sampled at a number of locations in the basin. By making the assumption that the concentration sampled each time at a particular place is drawn from the same water (i.e. that the data are statistically stationary in time), an average observed concentration can be computed for each sample point. The observed average concentrations can be compared with the expected mean annual concentrations computed from the flow accumulation model by drawing a circle at each sample point having an area proportional to the number of samples, and using a consistent color coding scheme for the expected and observed average concentrations. This analysis quickly shows where pollution is at or near expected levels, where point sources are raising the concentration well above the levels expected from non-point sources alone and provides a spatial picture of the variation of observed pollutant levels. Like the mean annual flow analysis, this grid-based non-point source pollution assessment method is simplified but it makes use of the observed hydrologic data and the GIS data normally available in a region in a reasonable way and yields useful results.

Time Varying Water Balance Models

A *water balance model* is a representation of the mass balance of water within a particular control volume. It is a physical statement of the law of conservation of mass which states that matter cannot be created or destroyed. As a result, the rate of change of storage of water within the control volume is equal to the difference between its rates of inflow and outflow across the control surface. One may distinguish in constructing a spatial hydrology model between the surface defining the outer boundary of the study region, and the surface defining the boundary of the spatial units within that region. A spatially distributed water balance model applies the law of conservation of mass to describe the mass balance within each spatial unit, and to this must be coupled a momentum equation (such as Darcy's law for groundwater flow) which defines how quickly water can move between units. Different sizes and shapes of spatial units are needed to deal with the different phases of the hydrologic cycle.

Atmospheric Water

Most water that falls on the land surface is derived from oceanic evaporation carried inland by atmospheric circulation, so it is appropriate to begin the study of hydrology by examining the motion of atmospheric water. The most useful way of doing this in a GIS context is to use the results of GCM modeling, where the acronym GCM means here General Circulation Model (this was the original meaning of this acronym before the more popular Global Climate Model came into vogue). In the United States, the National Meteorological Center in Maryland maintains a global GCM in continuous operation for numerical weather forecasting, which is updated each 12 hours with data from atmospheric soundings obtained from a global network of balloon-borne sensors released from weather stations, called the Global Data Assimilation System. The condition of the atmosphere (temperature, density, wind velocity, air pressure, moisture content) is calculated on a geographic grid of 2 degree cells covering the earth, using a very short time interval of the order of a few minutes, for a time horizon of a few days ahead. Each 12 hours, the forecasts are updated with new observations, and the process repeated. Summary statistics are available from the National Center for Atmospheric Research in Boulder, Colorado. A similar service is provided in Europe by the European Center for Medium Range Weather Forecasting in Reading, England.

The atmospheric water balance for a region, such as the State of Texas, can be computed by taking the vector field of atmospheric moisture flow, q, defined on the 2 degree grid, and interpolating the flow vectors onto points lying on the border of the State at regular intervals. The border of Texas is 7000 km long and points on the border at 100 km intervals seem to be appropriate. The lines joining these border points are also a set of vectors, L, which can be imagined to project vertically into the atmosphere and form a closed set of planes surrounding the atmosphere of the State, which are the control surface of an atmospheric control volume drawn over the State. The flow of water across any one of these planes can be found from the vector cross product:

$$Q = qxLy - qxLx$$

where the Easting and Northing components of the flow vector q = (qx, qy), and the border line vector L = (Lx, Ly) are cross multiplied. This is a rather interesting example of two different uses of the concept of a vector: firstly, in the sense of physical science for the atmospheric flow data, q; secondly, in the sense of GIS for the border line, L. By continuing this computation around the border, the net outflow of atmospheric moisture across the control surface can be determined, which is equal to difference between evaporation and precipitation on the land surface plus the change in atmospheric moisture content within the control volume. It turns out that the difference between atmospheric moisture inflow and outflow over Texas, Q, is usually less than 5% of the average atmospheric moisture flow over the State, so small errors in the flow computation can lead to substantial variations in the value of Q. The net evaporation from the land surface computed from this global data set may not be very accurate. The US National Meteorological Center is presently implementing a new mesoscale GCM over North America called the Eta model, using 40 km computational cells. The results of this model should permit more accurate atmospheric water balance computations to be made for regions like Texas.

Soil Water

Soil water is that water contained within the soil column, so the control volume for the water balance is a block of soil, and the computation consists in relating the change in soil moisture content to evaporation, precipitation, and outflow from the soil. At first sight, it would seem that the most appropriate spatial unit to use for a soil water balance would be a soil map unit, but these map units have very irregular shapes and a great range in size from one map unit to another. Climate data also play an important role and it appears that because the construction of spatially distributed climatic data usually involves interpolating climate onto a grid, that the grid cells used for that interpolation are also appropriate as "soil boxes" for constructing a soil water balance. Such climate cells are approximately 3' in size for the climate grids of the United States constructed by Daly et al., and for Africa constructed by Hutchinson et al. Willmott et al. constructed a global soil water balance model, and later Legates and Willmott developed a global climatology of monthly mean temperature and precipitation grids on 0.5 degree cells. Willmott uses the simple Thornthwaite soil water balance method, which requires only a single soil water parameter, the water holding capacity. Other soil water balance models, such as that in SWAT, have several soil layers and require more soil parameters. The recent emergence of a satellite derived net radiation balance of the earth provides net radiation estimates for the soil water balance, an important new data source.

The product of a soil water balance is a time history on a daily or a monthly basis of soil moisture content, evaporation and "water surplus" which is the water flowing from the soil to form surface runoff and groundwater recharge. Given the same input data, computation on a daily basis will always yield more water surplus than will computation on a monthly basis because daily precipitation is an episodic process, zero on most days, but when a precipitation event occurs, the soil moisture storage can be quickly filled up, thus producing a water surplus; if the same data are averaged over a monthly interval, it is as if the precipitation falls as a gentle mist, which may evaporate back to the atmosphere before the soil moisture capacity is filled. Interpolation of daily precipitation onto a grid is an uncertain undertaking because the spatial variation in daily precipitation is large. There is thus a challenge in constructing a GIS hydrology model for soil water balance in choosing the appropriate time interval for calculation.

Groundwater

There are two kinds of groundwater flow: *unconfined flow*, which occurs near the land surface and for which the water table or phreatic surface is the upper boundary of the saturated aquifer, and *confined flow* which occurs in deeper aquifers where the upper boundary on the flow is provided by an overlying aquitard, or hydrogeologic unit of low permeability. Unconfined flow is strongly influenced by surface hydrologic conditions, especially by seepage from streams crossing the region where the aquifer outcrops at the land surface. Confined flow is less influenced by surface conditions. Groundwater flow models usually use rectangular or triangular spatial units to represent elementary volumes of porous medium upon which the computations are to be performed. A GIS-based groundwater model can similarly be constructed using polygonal units provided the polygons are reasonably regularly shaped, but groundwater models constructed within GIS are not very computationally efficient.

In constructing a groundwater balance model, there are two computations to be performed: first, a water balance on each spatial unit in which all the inflows and outflows of the unit are used to determine the change in water storage and thus of the piezometric head within the unit; second, a flow computation between each pair of spatial units in which Darcy's law is used to determine the rate of groundwater flow as a function of the difference in head and the flow properties of the aquifer in the units. In a map-based groundwater modeling system, the first computation is done over all the polygons making up the aquifer, while the second is done over all the boundary lines of those polygons. Interaction between surface water in streams and underlying groundwater can be similarly determined by applying Darcy's law to the difference in piezometric head between the stream passing through an aquifer unit and the surrounding aquifer. All these computations need to be done on reasonably small units not more than say 20 km in cell size, because otherwise the head gradients in space become very small. Groundwater aquifers are usually quite confined in area and do not extend over the whole landscape, so unlike surface water flow which takes place everywhere, groundwater flow is more of a localized problem and a regional study needs to take into account each aquifer in the region individually, rather than considering groundwater flow to be a regional phenomenon.

Surface Water

Surface water is water in streams, lakes, wetlands and reservoirs. This water system is in some ways the most complex of all the phases of the hydrologic cycle, because it interacts with the other three phases, namely atmospheric water, soil water and groundwater, because the flow velocity is large compared to the velocity of groundwater flow, and because the flow environment is complicated, depending in part on the characteristics of the land surface and in part on the characteristics of the stream system. Fortunately, this is the area where GIS helps the most because of the detailed description of land surface features which can be presented in GIS. As described earlier, by making a suitable terrain analysis using DEM data, a conceptual model of the surface drainage system can be built up in which each watershed has one and only one stream draining it, and each watershed and stream pair can be assigned the same identification number. The watersheds so constructed are of two types: a source or head watershed in which the stream originates within the watershed, and an intermediate watershed where the stream flows both into and out of the watershed.

The stream network is manipulated so that each stream is represented by a single arc, and the arcs are flow ordered so that the from node is upstream and the to node is down stream. Each stream arc is enclosed within its associated watershed polygon. Watershed boundaries are delineated from each stream junction so at most a node can have two streams flowing into it and one flowing out of it. Three flow variables can be associated with each watershed: "From Flow", "To Flow", and "Polygon Flow". From Flow is that stream discharge at the from node; To Flow is the corresponding discharge at the to node; and Polygon Flow is that discharge which comes into the stream by drainage from the surrounding watershed. Polygon Flow is computed by applying a a unit hydrograph to the water surplus computed by the soil water balance model, and it may also include a component representing exchange of water between the stream and the underlying groundwater aquifer. This implies that the soil water surplus data may have to be spatially transferred from the soil water balance spatial units to the watershed units by using polygon overlay functions. The Polygon Flow can be divided by the length of the stream and added progressively to the discharge along the path of the stream, so in a time-averaged calculation, the discharge at a distance D from the upstream end of a stream of length L is given by:

$$\text{Discharge} = \text{From Flow} + (\text{Polygon Flow} / L) * D$$

In time-varying flow, the computation is more complex and stream routing methods such as the Muskingum method (Fread, 1993) are appropriate for computing the time distribution of the To Flow given the time distribution of the From Flow and the Polygon Flow. The time table structure is used to record the results of these calculations with a separate table being used for each of the three flow variables, a separate field for each watershed, and time on the vertical axis of the table.

In doing such computations, there is a choice between sequencing the computations "first in space and then in time" or "first in time and then in space", in other words, doing all the computations for a given watershed through time and then moving to the next watershed, or doing the computations for all watersheds for one time interval before moving to the next time interval. For regional hydrologic studies where the watersheds don't influence one another the "first in space then in time" method works best; for more localized hydraulic analyses where water level in the river influences groundwater levels, the simultaneous nature of the interactions makes the "first in time then in space" method necessary. The nature of the process interactions governs the computational sequence.

Water Utilization

All of the preceding discussion is valid for a pristine landscape untouched by human activity. But reservoir construction, pumping of water from rivers and groundwater systems, and discharge of wastewater all have profound effects on surface and groundwater flow and quality. Modeling the effects of water utilization permits GIS hydrology models to be useful as a basis for planning decisions on water facilities. Consider a pumping station withdrawing water from a river. In a spatial sense, this can be represented by a point on a river with attributes describing the time pattern of withdrawals. It is useful to locate this point by its relative distance from the upstream end of the reach (D/L in the nomenclature defined previously). This "proportional aliasing" provides a way of locating objects on river reaches which is somewhat independent of the scale of the map used to define the river reach spatially. River reaches are always longer on maps of larger scale but the relative location of a point remains reasonably stable across scales. Once the withdrawal is located, it can be included in the discharge computation.

CARTOGRAPHICAL MODELLING IN GIS

Cartographic Modelling is a generic way of expressing and organising the methods by which spatial variables, and spatial operations, are selected and used to develop an analytical solution with a GIS. Cartographic modelling is based on the concept of data layers, operations, and procedures. The purpose of the method is to create new map layers using existing map layers and operations that are sequenced in procedures.

A cartographical model is a graphic representation. It uses pictures linked by arrows in a flow chart. Its purpose is to help the analyst organize the necessary procedures as well as identify all the data needed for the study. It also serves as a source of documentation and reference for the analysis. In this way it can support communication between colleagues and contribute to the metadata that accompanies the outputs from spatial analysis.

Cartographic Models and GIS

Once you have created and inverted a cartographic model there is a four stage procedure that leads to the application of that model in a GIS:

Implementing a Cartographic Model

1. Identify the map layers or spatial data sets which are required.
2. Use logic and natural language to develop the process of moving from the available data to a solution.
3. Set up a flow chart with steps to graphically represent the above process. In the context of map algebra this flow chart represents a series of equations you must solve in order to produce the solution.
4. Annotate this flow chart with the commands necessary to perform these operations within the GIS you are using.

To explore these stages let us consider a supermarket siting example. We can complete stage one of the cartographic modelling process by identifying four data layers:

- Land_use
- Site_status
- River_map
- Roads_map

Stage two is completed by describing, in natural language, a scheme of spatial operations required to identify potential sites for the supermarket. Figure shows how stage three is completed by forming a flow chart to represent the logic in a GIS project. It is sometimes easier to visualise this with thumbnails of the data layers. Note how the appropriate verb (describing a spatial operation) has been added from table. The equation numbers, Eq1, Eq2,... Eq8, relate to equations in table.

Table: Presents four of the equations it would be necessary to solve as part of the process of finding a suitable site for the supermarket.

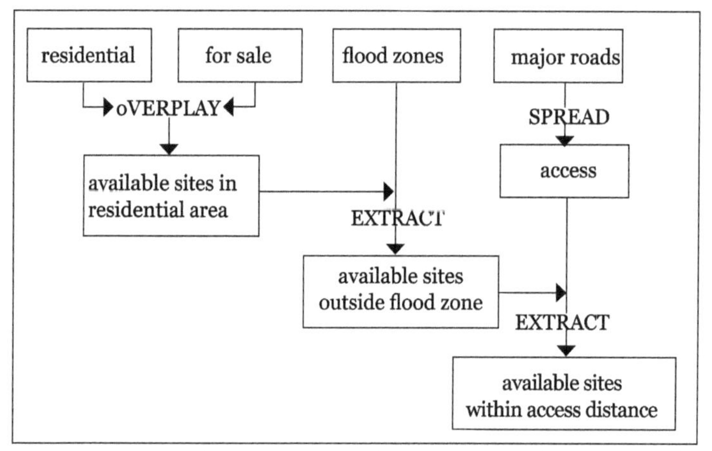

Flowchart of the operations needed to create a map identifying suitable locations for a supermarket.

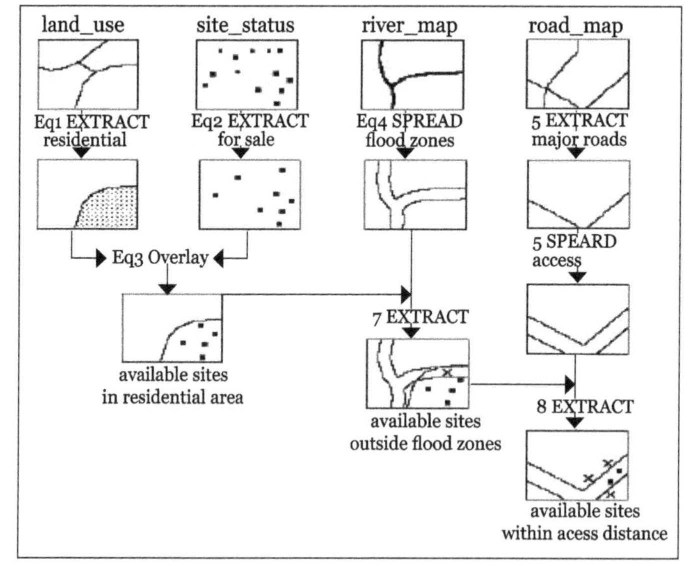

Flow chart with thumbnails.

Table: Algebraic equations from figure.

From LAND_USE 'extract' RESIDENTIAL a – b = c where: a = land_use map b = non residential zone c = residential
From SITE_STATUS ' extract' FOR_SALE d – e = f where: d = site_status map e = sites not for sale f = sites for sale
'Overlay' RESIDENTIAL and FOR_SALE Eq 3 c * f = g where: g = residential sites for sale

From the above exercise it should be apparent that the analytical power of cartographic modelling lies in the ability to combine a series of equations by using the results obtained from one equation as the input for the next. In this way a complex spatial problem can be tackled by breaking down the individual components of the problem into a series of smaller solvable equations.

The final stage in the modelling process is to annotate the flow chart with the appropriate commands from the GIS package in which it is intended to perform the analysis. In our example these commands are represented by verbs from Tomlin's Map Analysis Package (EXTRACT, OVERLAY and SPREAD). As was mentioned previously, however, different software applications translate these generic statements into their own terminology. Whilst making the running of training courses more profitable this localised terminology does nothing for spatial analysis and makes learning a new package more time consuming. These are trivial issues, however, compared to the benefit to be gained from implementing cartographic models.

Conditions for Applying Cartographic Models

There are two fundamental conditions required by any spatial analysis package: a consistent data structure and an iterative processing environment.

It has been a dream of mine for some time that I, as a data analyst, could forget about data structures, that I would perform spatial analysis without being concerned about whether a data layer was stored in a raster or vector format. Whilst there are many tools to convert between raster and vector representations it is still not possible to combine raster and vector data layers in a seamless analytical framework. This is because rasterisation and vectorisation are not clean, they might add something or change something in a data layer, and the effects of conversion are unpredictable. So, I am still required to prepare data in a consistent data structure. In general, there are more spatial operators coded to process raster data than vector data and so the raster data model may be preferred by data analysts.

The second condition, the iterative processing environment, logically sequences map analysis operations and involves the following:

1. Retrieving one or more map layer from a database,
2. Processing the data as specified by users,
3. Creating a new map containing the processing results,
4. Storing the new map for subsequent processing.

It is important that each new map is automatically georegistered to the other maps in the database. The output from one operation can then form the input to a later stage of processing. Such cyclical processing provides a flexible structure similar to "evaluating nested parentheticals" in traditional math. Within this structure, you first define the values for each variable and then solve the equation by performing the mathematical operations on the values in the order prescribed by the equation. For example, the equation for calculating percentage change in population starts with population at time 0 (X) and time 0+1 (Y). The difference is calculated by subtraction and then stored as an intermediate solution. The intermediate solution is divided by the initial population (X) to generate another intermediate solution that, in turn, is multiplied by 100 to calculate the solution (A) - the percentage change value.

Input values: X, Y

Intermediate solutions

Absolute change in valuc: $XY_{diff} = Y - X$

Proportionate change in value: $XY_{proportion} = XY_{diff} / X$

Solution: $A = XY_{proportion} \times 100\%$

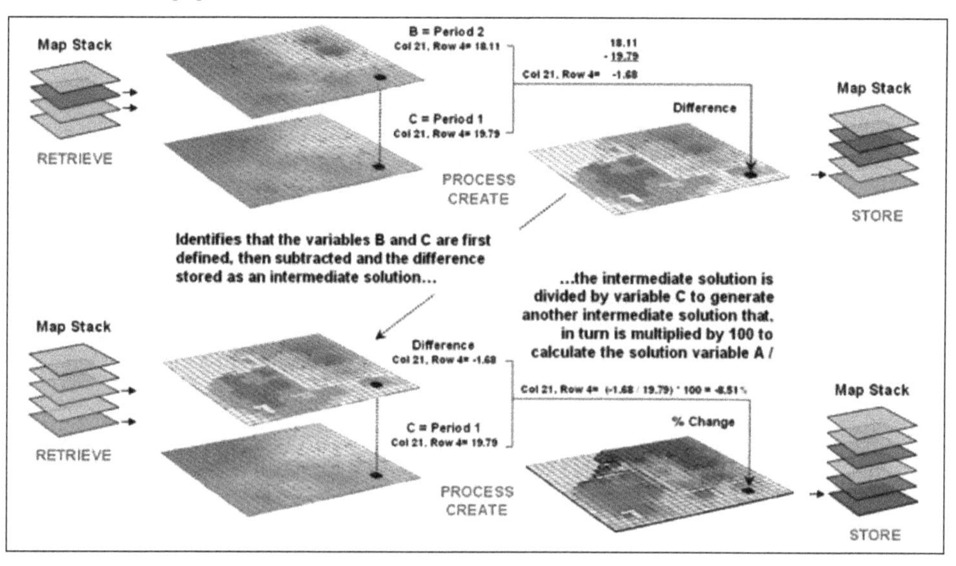

The same mathematical structure provides the framework for computer-assisted map analysis. The only difference is that the variables are represented by mapped data composed of thousands of georeferenced values. Figure shows a similar solution for calculating the pattern of percentage change in animal activitybut the calculations are performed for each grid cell in the study area. The result is a map that identifies the percent change at each location.

An iterative processing environment, analogous to basic math, is used to derive new map variables (red tones indicate decreased animal activity; green tones indicate increased activity; the example location shows a 8.51% decrease).

Map analysis quantifies the nature of change i.e. magnitude and direction of change in the thematic attribute, and the location of change using the spatial attribute. The characterization of "what" and "where" provides information needed for further environmental analysis, such as determining if areas of large increases in animal activity are correlated with particular cover types or near areas of low human activity.

Flexible Modelling in Practice

The mathematical structures and classification schemes form a conceptual framework that's easily adapted to modelling spatial relationships in physical and abstract systems. A major advantage is flexibility. For example, a model for siting a new highway can be developed as a series of processing steps. The analysis may consider economic and social concerns e.g. proximity to high housing density, visual exposure to houses, as well as purely engineering concerns e.g. steep slopes, water bodies. Combining physical and socio-economic concerns as part of an integrated spatial solution is another significant benefit. Furthermore, the ability to simulate various scenarios e.g. steepness is twice as important as visual exposure, and proximity to housing is four times more important than all other considerations, provides an opportunity to embed geospatial information into the decision-making process. By noting how often and where the proposed route changes as successive runs are made under varying assumptions, information on the unique sensitivity to siting a highway in a particular locale is described.

Compare this to a non-model based planning process. In the old environment, decision makers attempted to interpret results bounded by vague assumptions and system expressions of a specialist. Cartographic modelling, however, engages decision makers in an analytic process, because it documents the thought process and encourages interaction. It is the equivalent of a "spatial spreadsheet" that encapsulates the spatial reasoning of a problem and solves it using digital map variables.

MACHINE LEARNING IN GIS

Machine learning makes sense out of noisy data finding patterns that you'd never think existed. In other words, it's software that writes software.

Instead of applying a pre-built function, ML gains experience through repeated seen conditions and builds a model to apply in new situations.

For example, Google might use Bayesian classification to filter spam emails. Alternatively, Facebook might use it for facial recognition and automatically identify faces in images. And ML can even render Nicholas Cage in every movie ever made.

Types of Machine Learning

The two broad categories of machine learning are supervised and unsupervised. And they both can apply to GIS applications in various ways. First, what's the difference between the two?

Supervised Learning is just fitting data to a function for prediction. For example, if you plot millions of sample points in a graph, you can fit a line to approximate a function.

Unsupervised Learning recognizes what the data is using patterns from unlabelled data. For example, it takes millions of images and runs them through a training algorithm. After trillions of linear algebra operations, it can take a new picture and segment it into clusters.

Most importantly, machine learning is about optimally solving a problem. So it automatically learns on its own and improves from experience.

Lately, GIS is applying artificial intelligence into areas such as classification, prediction and segmentation.

Image Classification (Support Vector Machine)

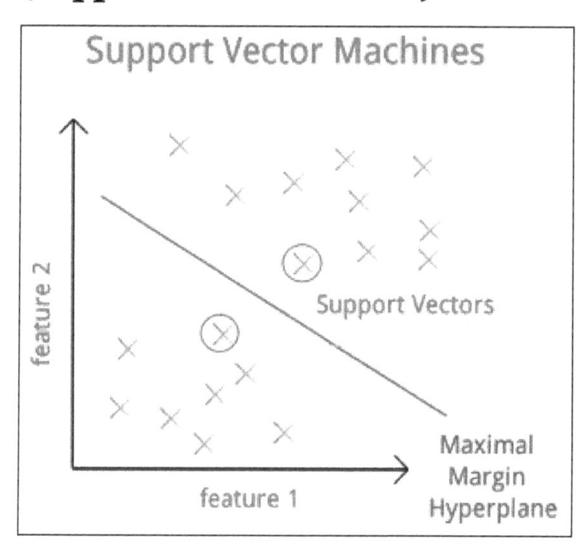

Support Vector Machine (SVM) is a machine learning technique that takes classified data and looks at the extremes. Next, it draws a decision boundary line based on the data called a "hyperplane". And the data points that the "hyperplane" margin pushes up against are the "support vectors".

And "support vectors" are what's important because they are the data points that are closest to the opposing classes. Because these points are the only ones considered, all other training points can be ignored in the model. Essentially, you feed SVM training samples of trees and grass. Based on this training data, it builds the model generating a decision boundary of its own.

Now, the results of this supervised classification aren't perfect and algorithms still have a lot more learning to do. We still need work on features like roads, wetlands and buildings. As algorithms get more training data, it will eventually improve to classify anywhere.

Prediction using Empirical Bayesian Kriging (EBK)

As you may know, kriging interpolation predicts unknown values based on spatial pattern. It estimates weights based on the variogram. And quality of the estimate surface is reflected in the quality of the weights. More specifically, you want weights that give an unbiased prediction and the smallest variance.

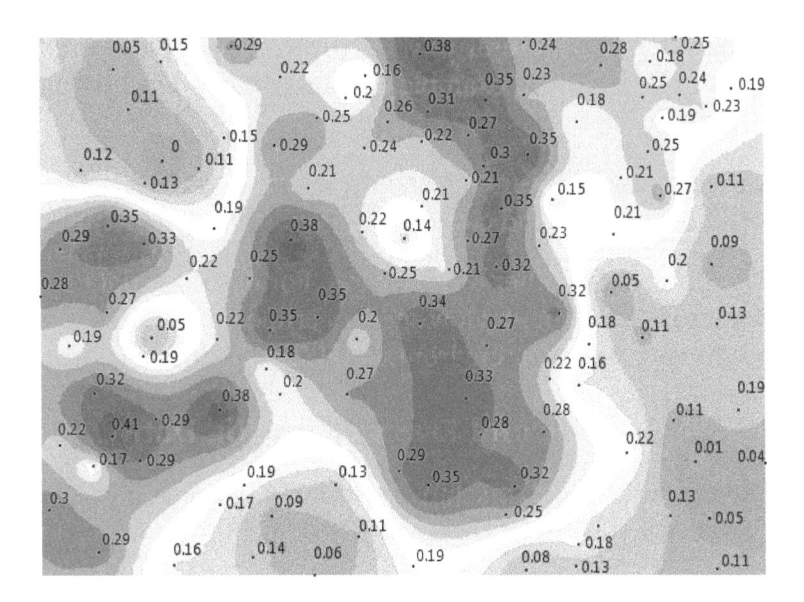

Unlike kriging that fits one whole model for an entire data set, EBK kriging simulates at least one hundred local models by sub-setting the whole data set. Because the model can morph itself locally to fit each individual semi-variogram using kriging methodology, it overcomes the challenge of stationarity.

In Empirical Bayesian Kriging (EBK), it predicts over and over again using a variety of simulations up to a hundred times. Each semi-variogram varies from each other. In the end, it mixes all of the semi-variograms for a final surface. You can't customize as you can with traditional kriging.

Finally, it outputs what it thinks is the best solution. Like a Monte Carlo analysis, it runs it repeatedly for you in the background. If it's a random process, you let the random process run out over a thousand times. You see the trends in the resulting data and use that to justify your selection. This is why EBK almost always predicts better than straight kriging.

Image Segmentation and Clustering with K-means

By far, the K-means algorithm is one of the most popular methods of clustering data. In K-means segmentation, it groups unlabeled data into the number of groups represented by the variable K.

This unsupervised learning approach iteratively assigns each data point into one of the K groupings based on similarity of features. For example, similarity can be based on spectral characteristics and location.

In an unsupervised classification, the k-means algorithm first segments the image for further analysis. Next, each cluster is assigned a land cover class.

However, GIS can use clustering in other unique ways. For example, data points could represent crime and you may want to cluster hot and low spots of crime. Alternatively, you may want to segment based on socioeconomic, health or environmental (like pollution) characteristics.

The Process of Deep Learning and Training for Big Data

Whether you're in GIS or another field, machine learning is all the buzz these days. It's about distilling big data sets. Because if you can let the computer detect the features, it will show you things you have never noticed.

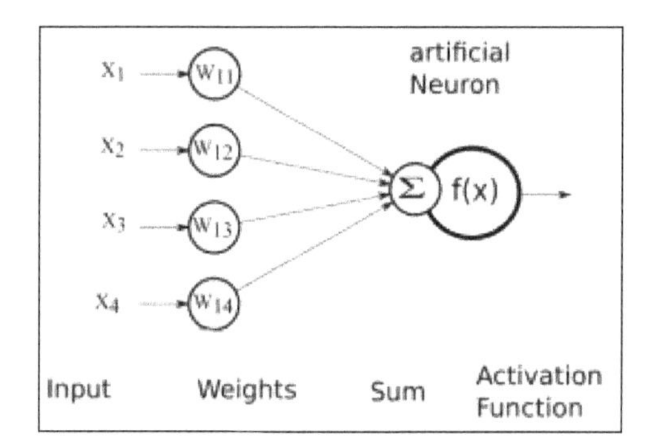

Because there's too much data, you can uncover inherent patterns from it. And the end result is a trained neural network with just a set of weighted values. When you train big data, this is when you're going to need all the firepower you can get. But once you have the model trained, it's just a model with a set of weights in a file and this why machine learning is a form of artificial intelligence – because you can train your data and then apply it to something entirely new and predict what it is. Overall, GIS uses machine learning for prediction, classification and clustering. AI and ML is still a growing field with a lot of framework still being developed daily.

SPATIAL JOIN TOOL IN GIS

The spatial join tool inserts the columns from one feature table to another based on location or proximity. Take a set of land parcels. Each land parcel has a point inside of it. By running a spatial join, you can transfer the point table columns into the land parcel layer. So that means that if the points have the owner name, this field gets transferred to the land parcels.

Spatial Join Types

The spatial join tool uses geographic proximity to combine attributes. It moves the columns from the join features to the target features.

Before you run a spatial join, you have to select a spatial join type. That is, the proximity the spatial join will search for.

Here are the most common spatial join types:

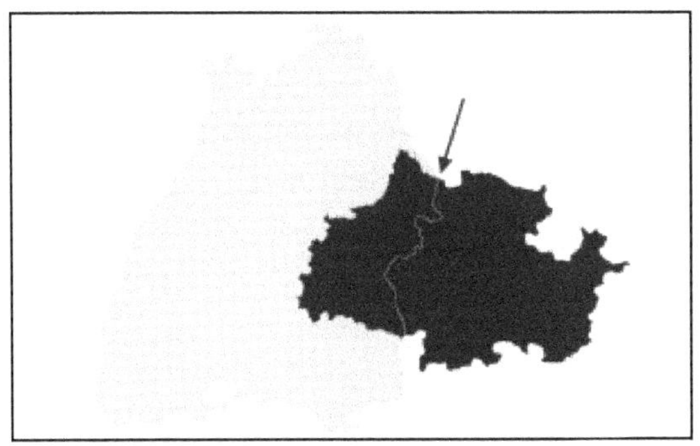

Intersect
Two features touch at any location.

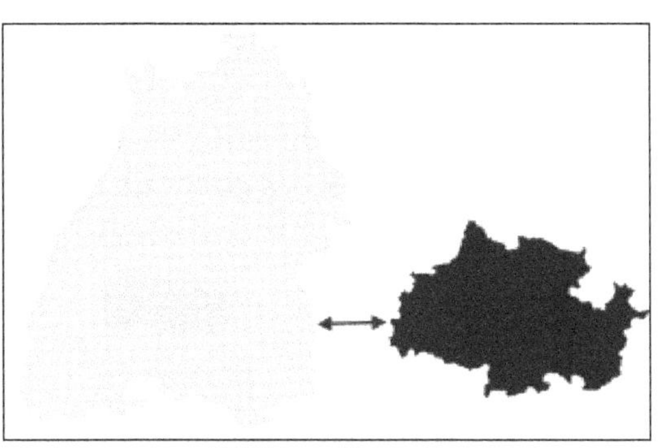

Within a distance
Two features are within a set distance.

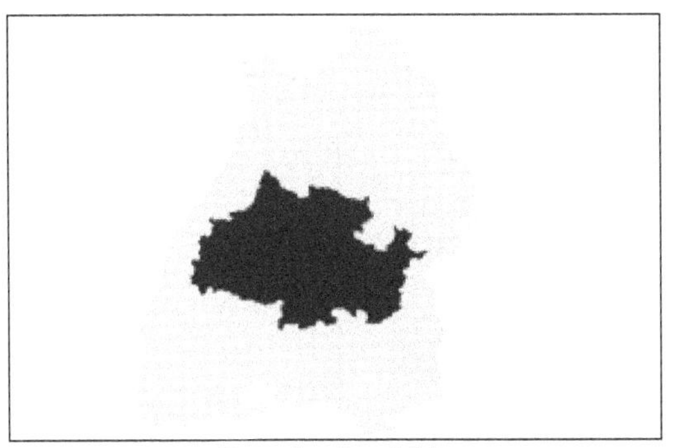

Completely within
The join feature is within the target feature.

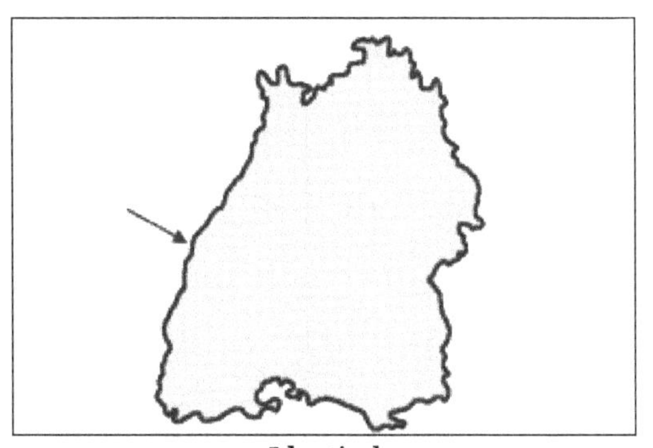

Identical
Both features match identically.

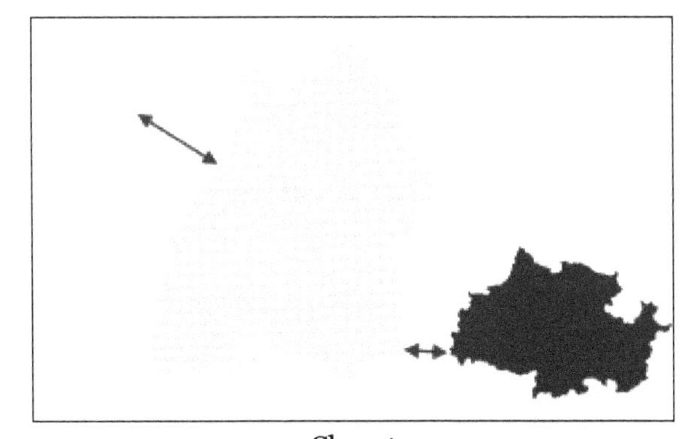

Closest
The join feature is closest to the target feature.

Depending on the spatial join type, it can affix one or several features into the target layer. In this case, you have to pick a join operation which include one-to-one or one-to-many.

Spatial Join Operations

Let's say you have multiple features that you want to combine into a target feature. In this case, you'll have to specify that it's a one-to-many operation.

A one-to-one operation will join a single feature from the join features into the target features. This is usually the first record the spatial join tool finds.

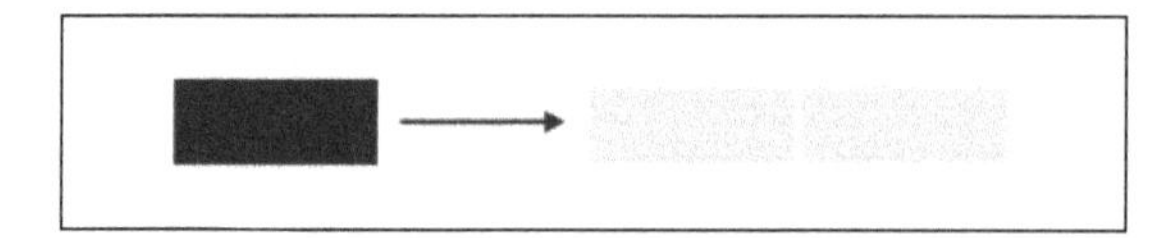

But a one-to-many operation will join all the features. It does so by creating multiple overlapping records. Each duplicate record contains the records from the joining features.

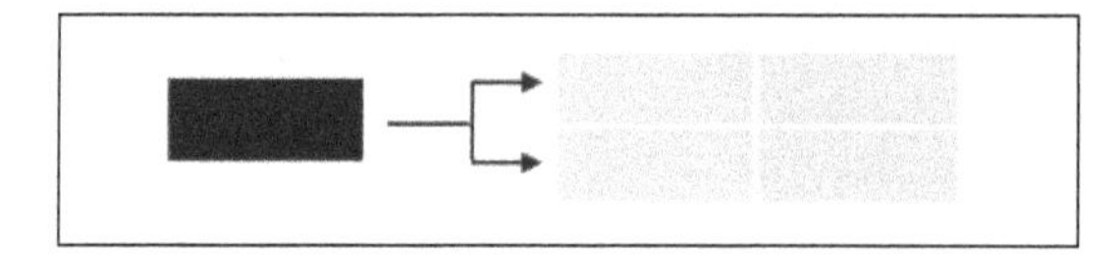

For example, if you have 1 land parcel. This parcel has 3 points in it with 3 different owners. A spatial join will create 3 identical land parcels. But each record will have the land owners name in it.

Spatial Join Examples

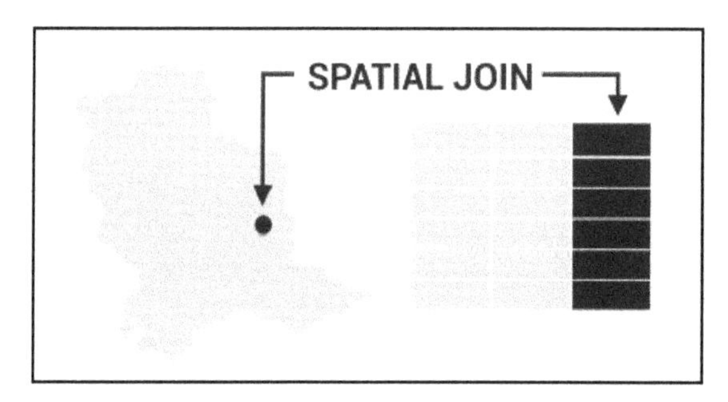

Spatial join is one of the top 7 geoprocessing tools in GIS. It's just as common as clipping, buffering, dissolving and merging.

Here are some examples of practical uses of applying a spatial join:

1. Students (points) reside in a school district (polygon). A spatial join will transfer the school district name to each student if they are within the polygon.
2. Well logs (points) are within a specific watershed (polygon). By running a spatial join, you can affix the watershed attributes to each well location.
3. Every county (polygon) is responsible for maintaining their own roads (lines). By running a spatial join, each road segment will add a column for the county it's in.

GIS SOFTWARE

A Geographic Information System (GIS Software) is designed to store, retrieve, manage, display, and analyze all types of geographic and spatial data. GIS software lets you produce maps and other graphic displays of geographic information for analysis and presentation.

What is GIS Mapping Software?

GIS software lets you produce maps and other graphic displays of geographic information for analysis and presentation. With these capabilities a GIS is a valuable tool to visualize spatial data or to build decision support systems for use in your organization.

A GIS stores data on geographical features and their characteristics. The features are typically classified as points, lines, or areas, or as raster images. On a map city data could be stored as points, road data could be stored as lines, and boundaries could be stored as areas, while aerial photos or scanned maps could be stored as raster images.

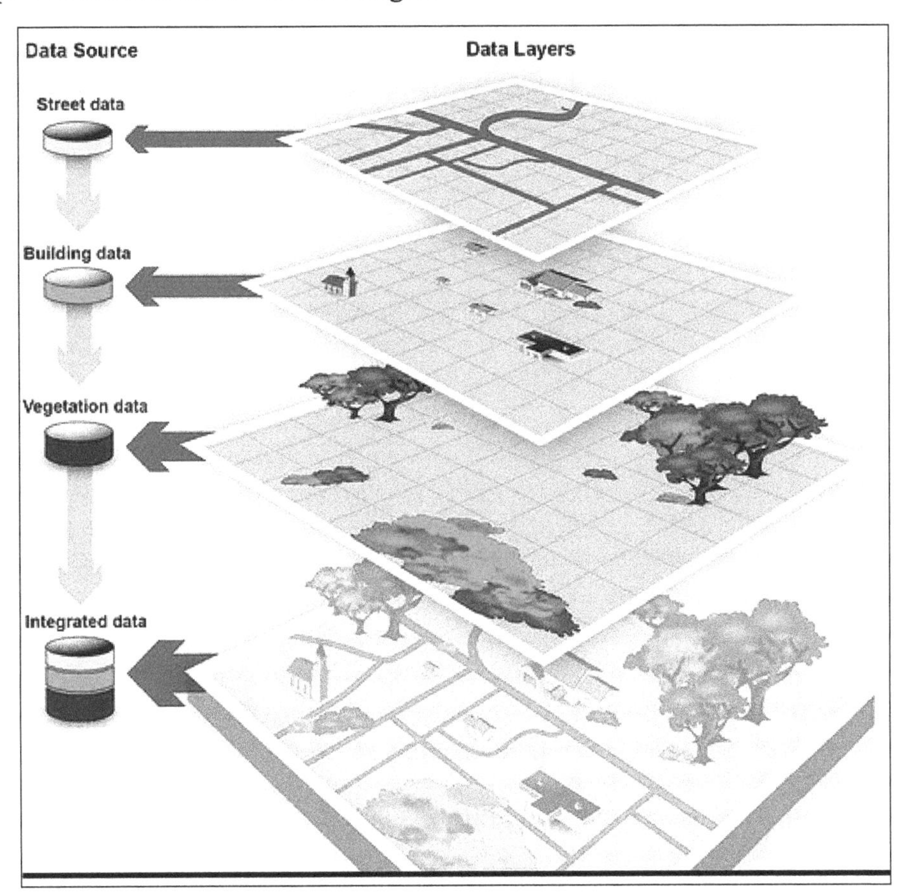

Geographic Information Systems store information using spatial indices that make it possible to identify the features located in any arbitrary region of a map. For example, a GIS can quickly identify and map all of the locations within a specified radius of a point, or all of the streets that run through a territory.

In addition to the above capabilities, Maptitude implements a professional-strength relational database, a feature critical for GIS software. Attribute data may be freely joined to and detached from geographic layers and tables. Relational data manipulation is integrated with robust and powerful geoprocessing for spatial queries, polygon overlay, and other location-based analyzes. This is supported seamlessly so that data are moved easily to and from relational tables and geographic databases. In addition, the Maptitude fixed-format binary table supports 32,767 fields and 1 billion records, and has unlimited character field widths.

Geographic Information System Software Features

Maptitude is one of the most popular GIS software packages, and has extensive functionality. A list of typical GIS capabilities is presented below, and these are available in Maptitude:

Maps and Layers

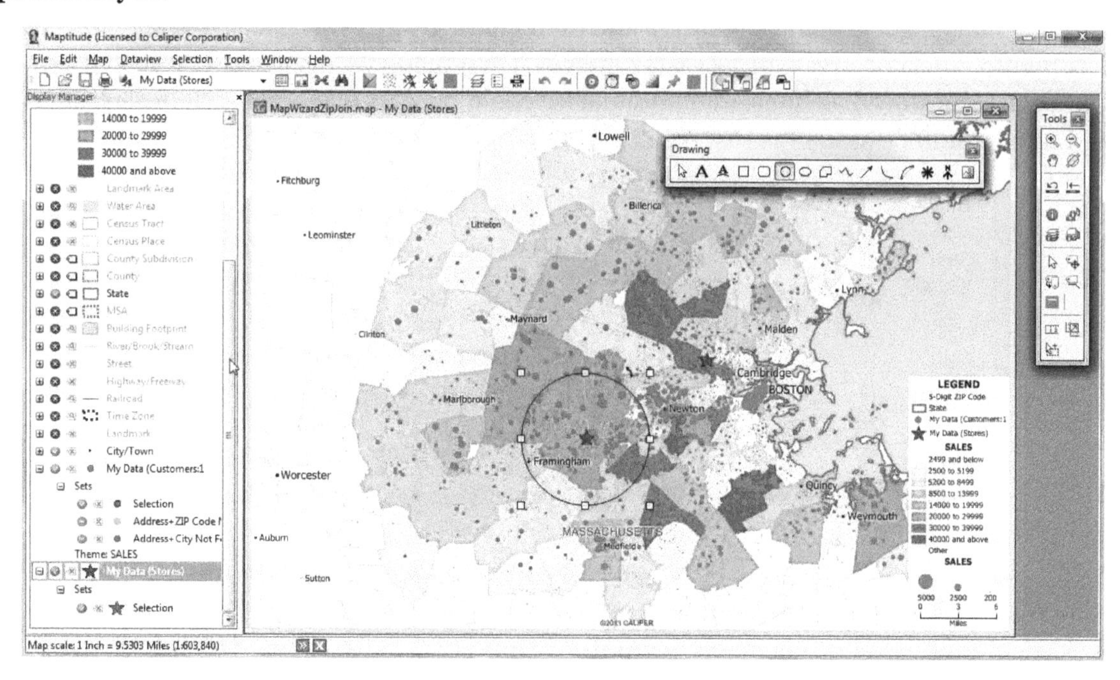

- The Create-a-Map Wizard allows users to easily create presentation-ready maps using their own data or the default maps.
- The Display Manager allows a map to be customized on-the-fly.
- User-defined preferences for map units, left/right side-of-road routing, file permissions, geocoding parameters, and many other settings.
- Toolbox and mouse-based map navigation is supported and includes panning, zooming, and magnifying.
- Map bookmark management allows the retrieval of custom map views.
- Multi-layer map feature query tools allow direct interrogation of spatial locations.
- A map librarian/manager allows the organization of various saved maps and comes with a library of pre-styled demographic maps.
- Geographic database layering controls allow customization of layer visibility and drawing order.
- Multiple maps can be open simultaneously, and can be duplicated, combined, synchronized, tiled, cascaded, and minimized/maximized.
- There is explicit map scale control including undo.

- Layer autoscaling allows customization of the scale at which layers are visible.
- An interactive map overview window provides perspective as you work and the ability to zoom anywhere in the study region.

Visualization

- Extensive layer style control includes font/style/opacity settings for points/lines/areas/labels/legends/drawings; point and area styles can use most image formats and their resolution can be controlled via scaling.
- Thematic visualizations include color, pattern/icon, dot-density, chart, scaled-symbol, and 3D prism themes.
- A drawing toolbox is provided, the drawing items are customizable, and there is a selection of north-oriented arrows.
- Each map has an editable legend that automatically lists displayed features and has a live scale bar.
- Stand-alone charting capabilities include pie, bar, line, area, scatter, and function charts.
- Advanced text label placement and management tools include live label manipulation en-masse or individually, automated positioning, callouts/rotation, font control, multi-line, framing, hiding, styling, prioritizing, stretching, spacing, autoscaling, and additional text manipulation settings.
- Maps and graphics can be copy/pasted or saved as pictures/bitmaps (with optional quality/resolution settings) for insertion into MS Office and other external applications.
- Printing to any printer/paper size is supported, with a wide variety of spatial print options including using fixed scale, with actual point sizes, and as pre-rendered images.
- Report/layout creation can utilize settings for snap grids, rulers, paper size/orientation, dimensions, margins, alignment, print options, automated district printing, and a variety of other graphics software oriented options.
- Map interaction can be recorded to video.
- Layer style/label/autoscale override is provided through the Feature Display tool.
- Cartographic coloring uses Brelaz's Dsatur algorithm to assign colors that ensure that no two adjacent regions have the same color.

Geocoding

- The tabular and geographic find tool can identify locations anywhere on earth.
- Robust and flexible pin-mapping tools support geocoding by address, postal code, city/town, join, coordinate, longitude/latitude, by any populated place in the world (village, town, city), and also manually.

- Custom geocodable indexes can be created to pin-map based on external datasets.
- Geotagged images from smart phones, tablets, or GPS-enabled devices can be mapped.

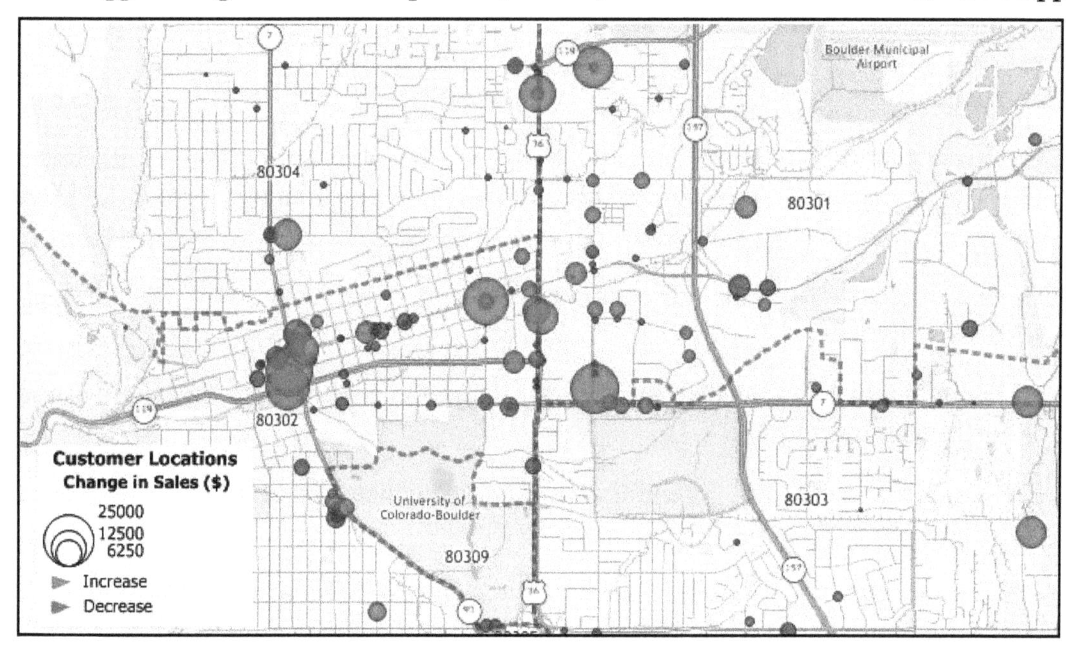

GIS Mapping Tools and Geographic Analysis

Geographic analysis tools are the most valuable component of GIS software because they let you analyze the geographic components of your data. Below are some of the geographic analysis tools that are standard in Maptitude:

Territory Building Tools: Districts/Territories can be created using map-based filters or via tabular groupings.

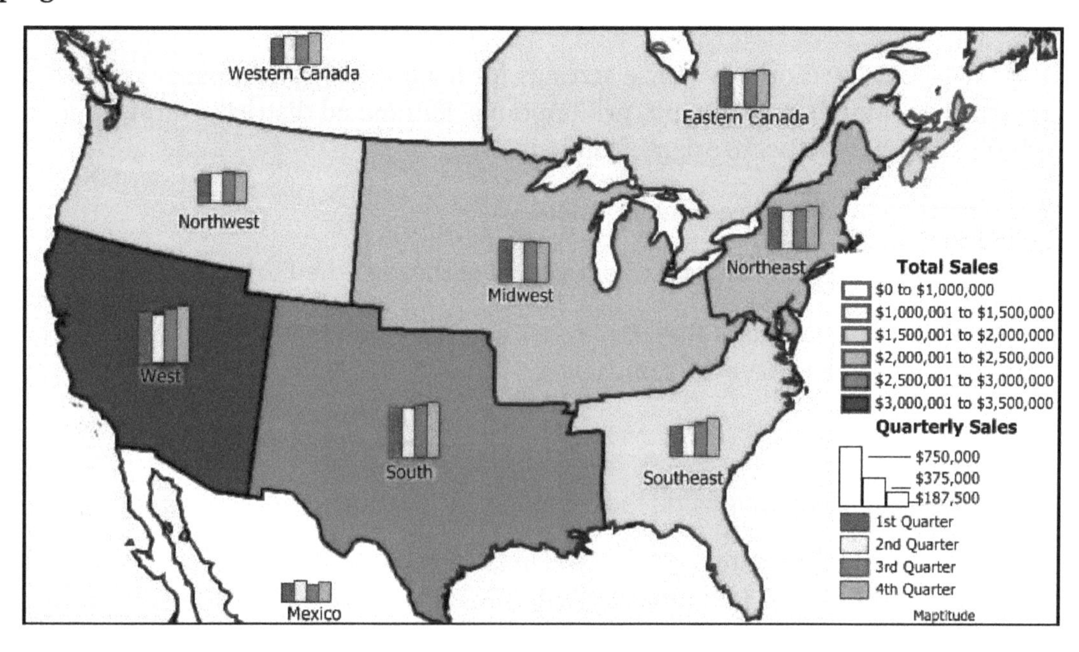

Buffers: Circular buffers/bands for analyzing proximity.

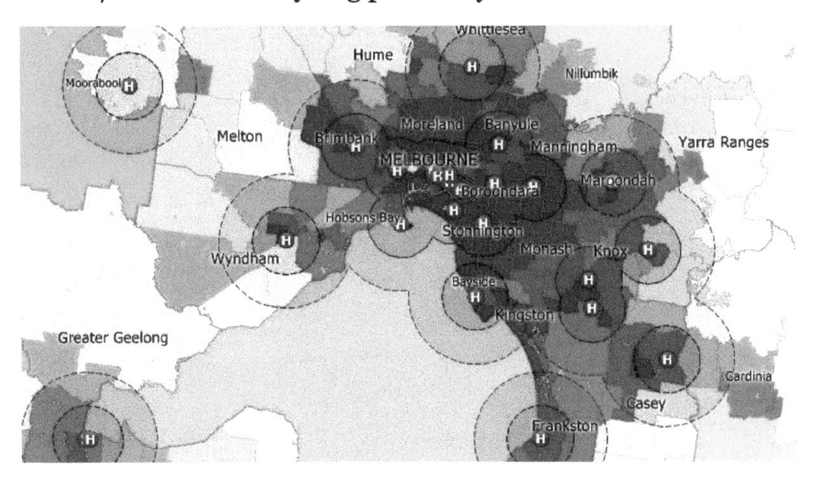

Facility Location: A facility location tool identifies the best location for one or more facilities from a set of candidate sites.

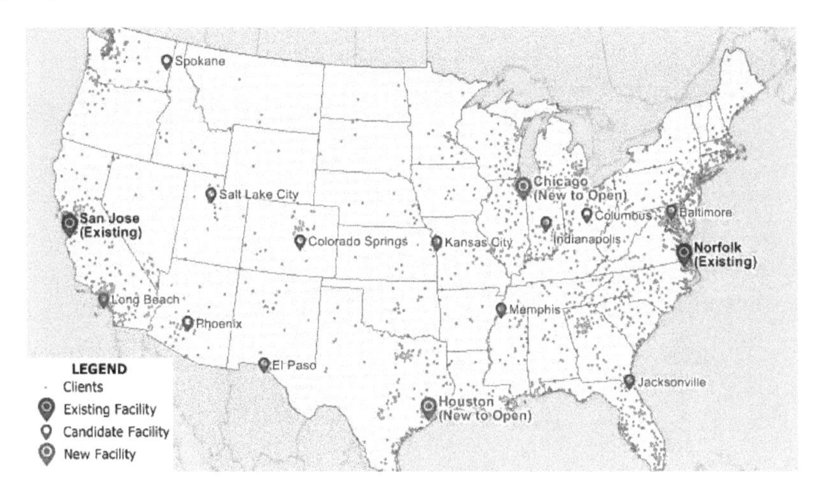

Geographic Overlay: Geographic overlay/aggregation is supported and allows attribute assignment between layers based on percentage overlap for estimating demographics of territories, buffers, areas of influence, and more.

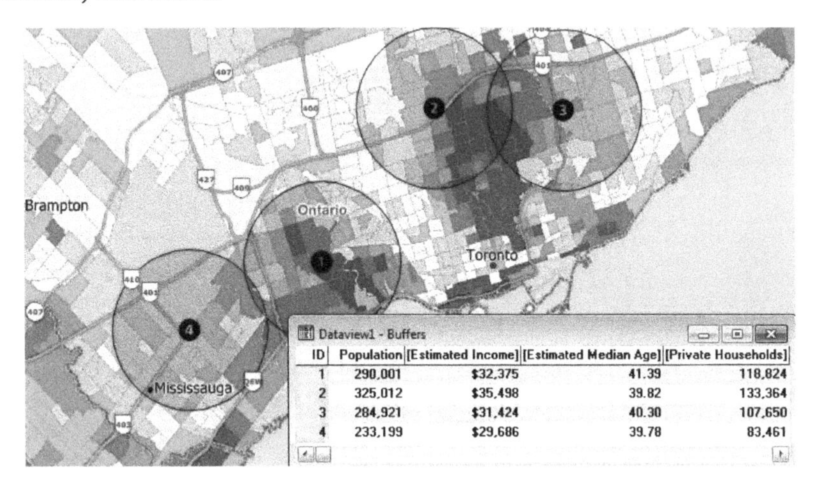

Dataview1 - Buffers

ID	Population	[Estimated Income]	[Estimated Median Age]	[Private Households]
1	290,001	$32,375	41.39	118,824
2	325,012	$35,498	39.82	133,364
3	284,921	$31,424	40.30	107,650
4	233,199	$29,686	39.78	83,461

Hot Spots: Kernel-based density grids can be created using the quartic, triangular, uniform, or count methods, and allow "hot-spot" mapping.

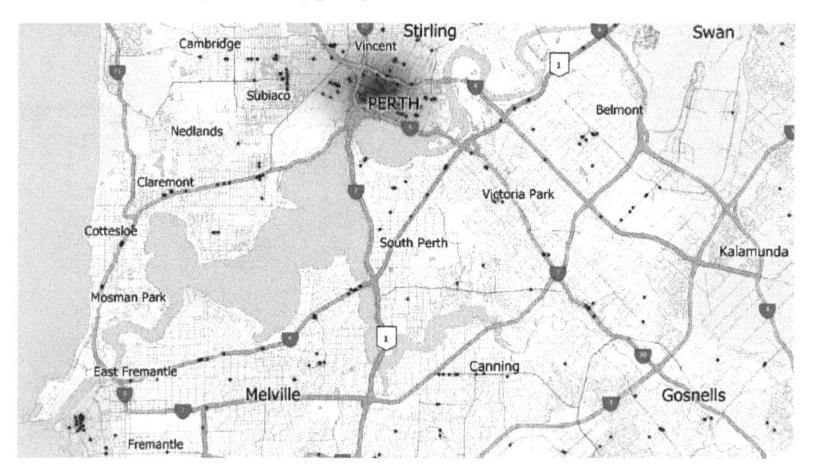

Weighted Center: Weighted center calculations allow the identification of centers of "gravity" among points.

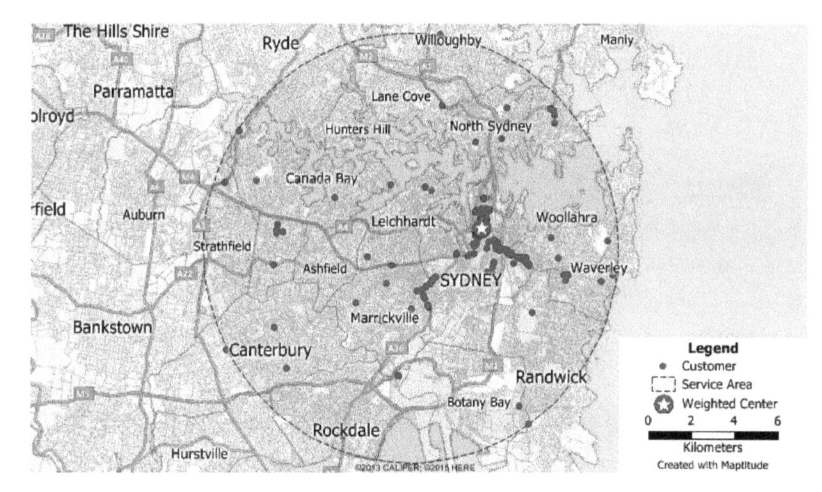

Shortest Path: The shortest path calculations allow for minimizing the cost of the path as an ordered/unordered route with options to produce directions and to return to the origin.

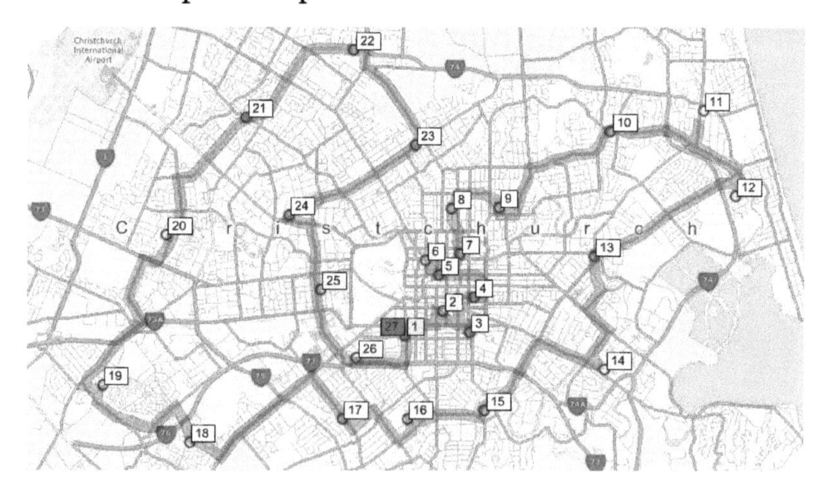

Drive-time Bands: Drive time/distance bands allow you to visualize the extent to which locations can be accessed within a certain drive time or distance.

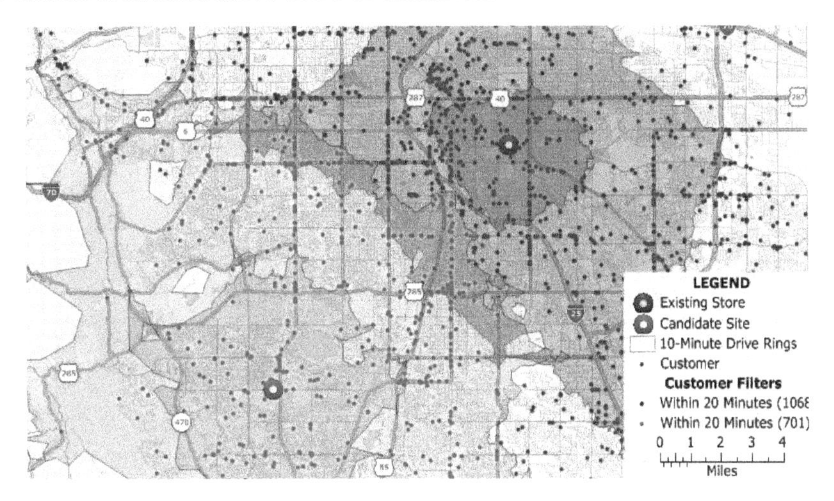

Rive-time Territories: Drive time partitions allow regions across a line layer to be defined based on network cost.

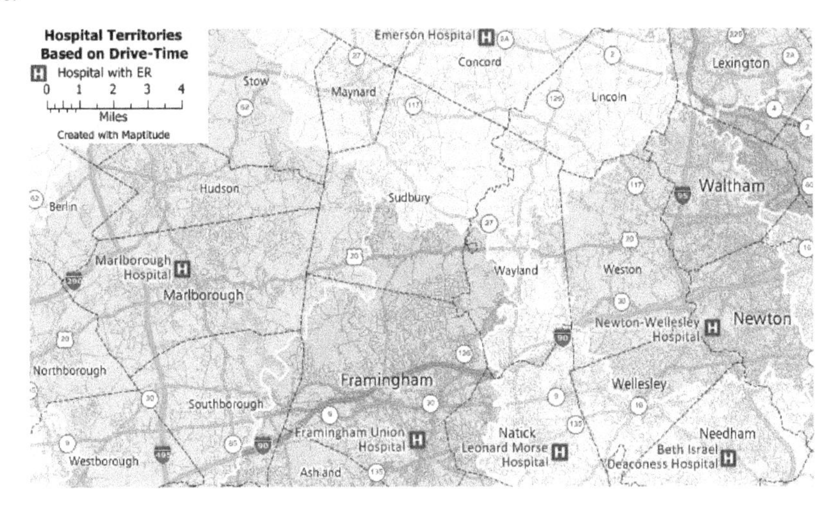

Measuring Tools: Length/area measurement tools allow map-based calculations.

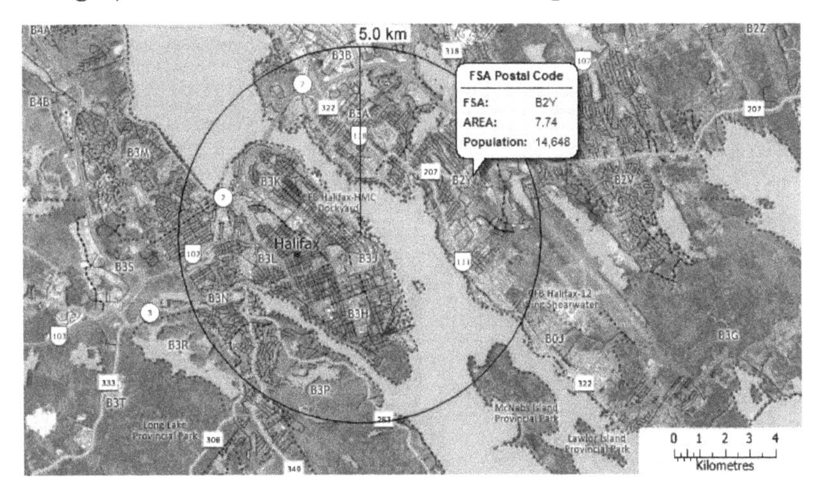

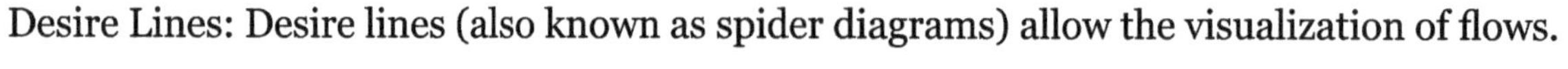

Desire Lines: Desire lines (also known as spider diagrams) allow the visualization of flows.

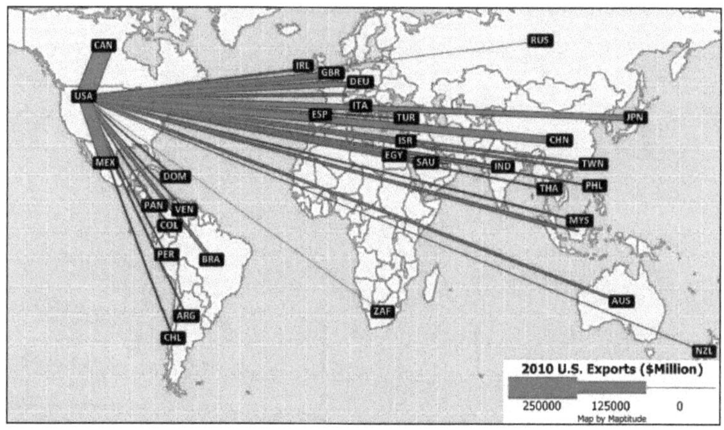

Surface Analysis: Surface analysis tools include spot height data querying, surface profiling, viewsheds, contouring, 3D terrain visualization, DEM/TIN creation, and the calculation of terrain shortest paths.

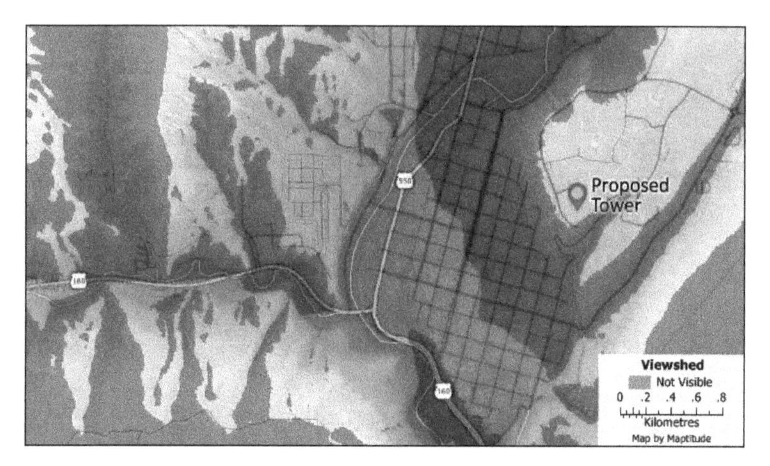

Data Classification: Data classification methods include: quantiles, equal weight, equal interval, standard deviation, nested means, arithmetic or geometric progression, Fisher-Jenks/optimal breaks, categories, and manual classification (by range, counts, or percentages).

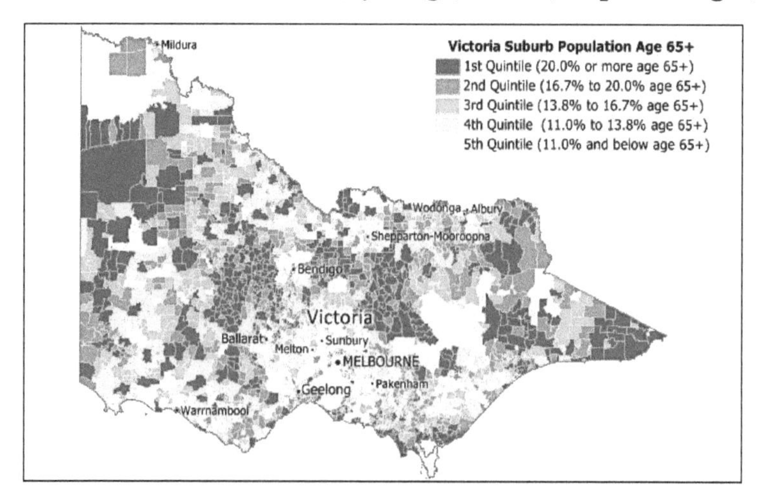

Areas of Influence: Areas-of-Influence are a powerful GIS tool that divide the study area using a triangulated irregular network (TIN).

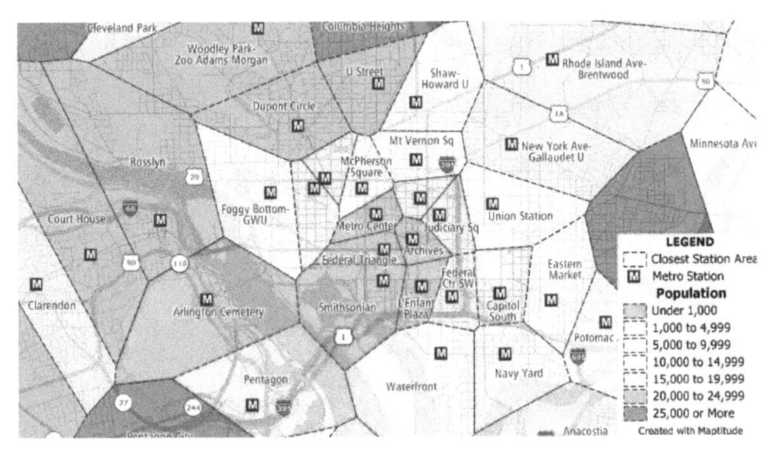

GPS Support: GPS support includes the ability to read/animate/import GPS data, overlay tracks with aerial photos and topographic or vector maps, track real time GPS locations, create vector line/point layers from GPS playback files, and import/export formats such as GPX (the GPS Exchange Format)

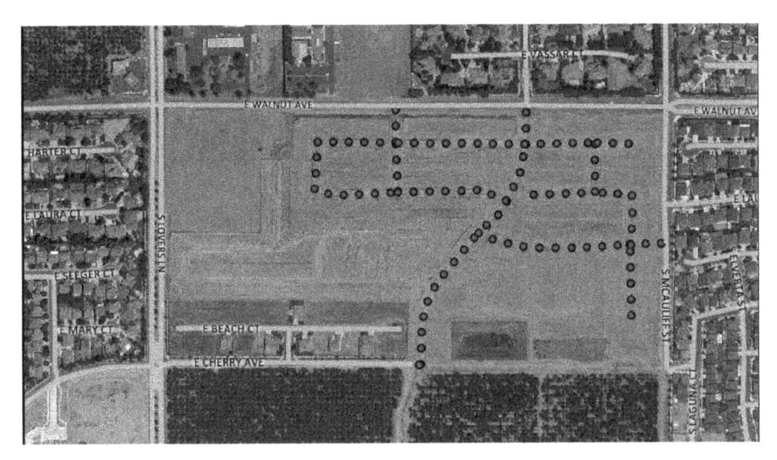

Spatial Queries: Filter features based on geographic location, proximity to other features, by radius, by pointing, by polygon, or based on a value or condition.

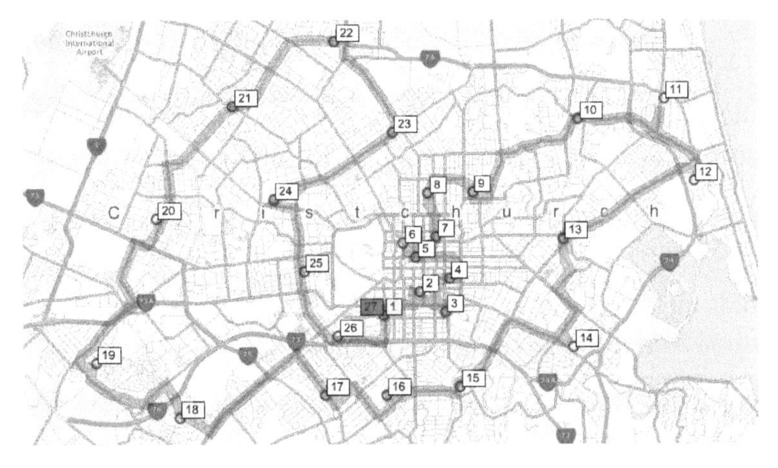

Internet Mapping: Map server products such as Maptitude for the Web and Cloud and SaaS location-based applications allow you to share your geographic data as device independent and mobile-friendly interactive maps. You can also add mapping functions to your web site or web-based solutions, such as providing the public with access to assessor parcel maps and valuation data.

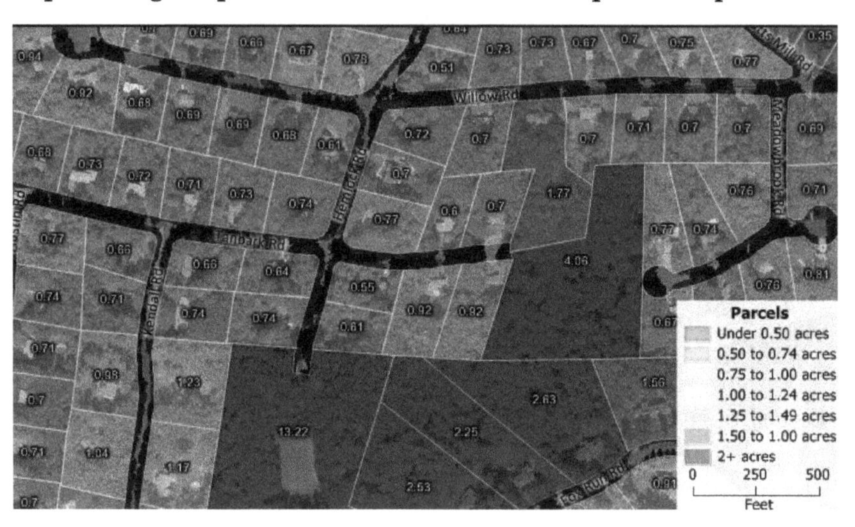

Imagery

- Image layer and aerial photo tools include registration, a manager/librarian, contrast control, smoothing (from 2x2 to 10x10) and interpolation (nearest neighbor, bilinear, high quality bilinear, bicubic, high quality bicubic).
- The image servers supported are Google Earth and OGC Web Map Services (WMS).

Database

- The Maptitude GIS program has a powerful proprietary relational database.
- Support is provided for over 50 file types and more than 100 GIS and CAD formats, some natively including Excel, MS Access, ODBC, dBase, CSV, ASCII, ArcGIS platform formats (Esri Shapefile and Personal Geodatabase), MapInfo TAB, Oracle Spatial, and SQL Server Spatial
- Support is provided for exporting to many formats including Excel, dBase, CSV, ASCII, Lotus, Google KML, ArcGIS platform formats (Esri Shapefile and ArcMap Document), MapInfo MIF, Oracle Spatial, SQL Server Spatial and AutoCAD DXF.
- Table tools include the ability to transpose, group/aggregate, identify duplicates, calculate statistics, convert longitude/latitude to XY coordinates, print mailing labels, copy/paste values, and perform undo/redo of edits.
- Regression and binary logit models can be estimated on any map layer or table.
- Table field tools include the ability to hide, show, filter, lock, format, multi-field sort, create live expression/formula fields, and perform multi-cell fills.
- Database modify tools include the ability to add/delete records/fields, delete filtered records, set aggregation rules, apply look-up table coding, and define field header balloon pop-up text.

- Database joins can be aggregate/non-aggregate and as one-to-one, one-to-many, or many-to-one joins.
- Multiple filters per layer or database can be created using SQL type queries, spatial queries (coincident, adjacent, within, and many more), and data classification methods.
- Topological/non-topological spatial databases can be created for points, lines, areas, or grids.
- Topological/non-topological layer (line/point/area) editing tools include the ability to use digitizers, create one-way streets, copy and paste lines, merge/split features/attributes, add/delete/move features, line/area conversion, point-to-line conversion, merging layers, clipping/masking geography by region/area, and undo/redo of edits.
- There is comprehensive projection, datum, and coordinate system support both natively and via import/export, and this operates in conjunction with tools such as vector rubber-sheeting and on-the-fly raster layer reprojection.
- Any record can be linked to multiple files including photos, documents, web pages, and slide-shows.

Development Platform

- The Geographic Information System Developer's Kit (GISDK) has 850+ Caliper Script functions that can be called to create add-ins, build custom applications, and to access Maptitude from. NET or as a COM Object.

Processing

Maptitude supports the latest Windows operating systems, file types, and common design elements. Maptitude runs as a 32-bit or 64-bit application on 32-bit or 64-bit Windows. Advantages of a 64-bit Maptitude include:

- Save to much higher resolution images.
- Use more memory than the previous 4GB 32-bit limit.
- Open/import files via 64-bit Microsoft Office (e.g. Excel and Access).

ARCGIS

ArcGIS is a geographic information system (GIS) for working with maps and geographic information. It is used for creating and using maps, compiling geographic data, analyzing mapped information, sharing and discovering geographic information, using maps and geographic information in a range of applications, and managing geographic information in a database.

The system provides an infrastructure for making maps and geographic information available throughout an organization, across a community, and openly on the Web.

ArcGIS includes the following Windows desktop software:

- ArcReader, which allows one to view and query maps created with the other ArcGIS products.
- ArcGIS for Desktop, which is licensed under three functionality levels:
    - ArcGIS for Desktop Basic (formerly known as ArcView), which allows one to view spatial data, create layered maps, and perform basic spatial analysis.
    - ArcGIS for Desktop Standard (formerly known as ArcEditor), which in addition to the functionality of ArcView, includes more advanced tools for manipulation of shapefiles and geodatabases.
    - ArcGIS for Desktop Advanced (formerly known as ArcInfo), which includes capabilities for data manipulation, editing, and analysis.

There are also server-based ArcGIS products, as well as ArcGIS products for PDAs. Extensions can be purchased separately to increase the functionality of ArcGIS.

ArcGIS version history	
Version	Released
8.0	1999-12-27
8.0.1	2000-01-13
8.1	2001-05-01
8.2	2002-05-10
8.3	2003-02-10
9.0	2004-05-11
9.1	2005-05-25
9.2	2006-11-14
9.3	2008-06-25
9.3.1	2009-04-28
10.0	2010-06-29
10.1	2012-06-11
10.2	2013-07-30
10.2.1	2014-01-07
10.2.2	2014-04-15
10.3	2014-12-10
10.3.1	2015-05-13
10.4	2016-02-18
10.4.1	2016-05-31
10.5	2016-12-15
10.5.1	2017-06-29
10.6	2018-01-17
10.6.1	2018-07-16
10.7	2019-03-21

Prior to the ArcGIS suite, Esri had focused its software development on the command line Arc/INFO workstation program and several Graphical User Interface-based products such as the ArcView GIS 3.x desktop program. Other Esri products included MapObjects, a programming library for developers, and ArcSDE as a relational database management system. The various products had branched out into multiple source trees and did not integrate well with one another. In January 1997, Esri decided to revamp its GIS software platform, creating a single integrated software architecture.

ArcMap 8.0

In late 1999, Esri released ArcMap 8.0, which ran on the Microsoft Windows operating system. ArcGIS combined the visual user-interface aspect of ArcView GIS 3.x interface with some of the power from the Arc/INFO version 7.2 workstation. This pairing resulted in a new software suite called ArcGIS including the command-line ArcInfo workstation (v8.0) and a new graphical user interface application called ArcMap (v8.0). This ArcMAP incorporating some of the functionality of ArcInfo with a more intuitive interface, as well as a file management application called ArcCatalog (v8.0). The release of the ArcMap constituted a major change in Esri's software offerings, aligning all their client and server products under one software architecture known as ArcGIS, developed using Microsoft Windows COM standards. While the interface and names of ArcMap 8.0 are similar to later versions of ArcGIS Desktop, they are different products. ArcGIS 8.1 replaced ArcMap 8.0 in the product line but was not an update to it.

ArcGIS Desktop 8.1 to 8.3

ArcGIS 8.1 was unveiled at the Esri International User Conference in 2000. ArcGIS 8.1 was officially released on April 24, 2001. This new application included three extensions: 3D Analyst, Spatial Analyst, and GeoStatistical Analyst. These three extension had become very powerful and popular in ArcView GIS 3.x product line. ArcGIS 8.1 also added the ability to access data online, directly from the Geography Network site or other ArcIMS map services. ArcGIS 8.3 was introduced in 2002, adding topology to geodatabases, which was a feature originally available only with ArcInfo coverages.

One major difference is the programming (scripting) languages available to customize or extend the software to suit particular user needs. In the transition to ArcGIS, Esri dropped support of its application-specific scripting languages, Avenue and the ARC Macro Language (AML), in favour of Visual Basic for Applications scripting and open access to ArcGIS components using the Microsoft COM standards. ArcGIS is designed to store data in a proprietary RDBMS format, known as geodatabase. ArcGIS 8.x introduced other new features, including on-the-fly map projections, and annotation in the database.

ArcGIS 9.x

ArcGIS 9 was released in May 2004, which included ArcGIS Server and ArcGIS Engine for developers. The ArcGIS 9 release includes a *geoprocessing* environment that allows execution of traditional GIS processing tools (such as clipping, overlay, and spatial analysis) interactively or from any scripting language that supports COM standards. Although the most popular of these is Python, others have been used, especially Perl and VBScript. ArcGIS 9 includes a visual programming environment, similar to ERDAS IMAGINE's Model Maker (released in 1994, v8.0.2). The Esri version is called ModelBuilder and as does the ERDAS IMAGINE version allows users to graphically link geoprocessing tools into new tools called *models*. These models can be executed directly or exported to scripting languages which can then execute in batch mode (launched from a command line), or they can undergo further editing to add branching or looping.

On June 26, 2008, Esri released ArcGIS 9.3. The new version of ArcGIS Desktop has new modeling tools and geostatistical error tracking features, while ArcGIS Server has improved performance, and support for role-based security. There also are new JavaScript APIs that can be used to create mashups, and integrated with either Google Maps or Microsoft Virtual Earth.

At the 2008 Esri Developers Summit, there was little emphasis on ArcIMS, except for one session on transitioning from ArcIMS to ArcGIS Server-based applications, indicating a change in focus for Esri with ArcGIS 9.3 for web-based mapping applications.

In May 2009, Esri released ArcGIS 9.3.1, which improved the performance of dynamic map publishing and introduced better sharing of geographic information.

ArcGIS 10.x

- In 2010 Esri announced that the prospective version 9.4 would become version 10 and would ship in the second quarter of 2010.
- In June 2012 Esri released ArcGIS 10.1.
- In July 2013 Esri released ArcGIS 10.2.
- In December 2014 Esri released ArcGIS 10.3. The release included ArcGIS Pro 1.0, which became available in January 2015.
- In February 2016 Esri released ArcGIS 10.4.
- In December 2016 Esri released ArcGIS 10.5.
- In January 2018 Esri released ArcGIS 10.6.
- In March 2019 Esri released ArcGIS 10.7.

Geodatabase

Older Esri products, including ArcView 3.x, worked with data in the shapefile format. ArcInfo Workstation handled coverages, which stored topology information about the spatial data. Coverages, which were introduced in 1981 when ArcInfo was first released, have limitations in how they handle types of features. Some features, such as roads with street intersections or overpasses and underpasses, should be handled differently from other types of features.

ArcGIS is built around a geodatabase, which uses an object-relational database approach for storing spatial data. A geodatabase is a "container" for holding datasets, tying together the spatial features with attributes. The geodatabase can also contain topology information, and can model behavior of features, such as road intersections, with rules on how features relate to one another. When working with geodatabases, it is important to understand feature classes which are a set of features, represented with points, lines, or polygons. With shapefiles, each file can only handle one type of feature. A geodatabase can store multiple feature classes or type of features within one file.

Geodatabases in ArcGIS can be stored in three different ways – as a “file geodatabase”, a “personal geodatabase”, or an “ArcSDE geodatabase”. Introduced at 9.2, the file geodatabase stores information in a folder named with a. gdb extension. The insides look similar to that of a coverage but is not, in fact, a coverage. Similar to the personal geodatabase, the file geodatabase only supports a single editor. However, unlike the personal geodatabase, there is virtually no size limit. By default, any single table cannot exceed 1TB, but this can be changed. Personal geodatabases store data in Microsoft Access files, using a BLOB field to store the geometry data. The OGR library is able to handle this file type, to convert it to other file formats. Database administration tasks for personal geodatabases, such as managing users and creating backups, can be done through ArcCatalog. Personal geodatabases, which are based on Microsoft Access, run only on Microsoft Windows and have a 2 gigabyte size limit. Enterprise (multi-user) level geodatabases are handled using ArcSDE, which interfaces with high-end DBMS such as PostgreSQL, Oracle, Microsoft SQL Server, DB2 and Informix to handle database management aspects, while ArcGIS deals with spatial data management. Enterprise level geodatabases support database replication, versioning and transaction management, and are cross-platform compatible, able to run on Linux, Windows, and Solaris.

Also released at 9.2 is the personal SDE database that operates with SQL Server Express. Personal SDE databases do not support multi-user editing, but do support versioning and disconnected editing. Microsoft limits SQL Server Express databases to 4GB.

ArcGIS for Desktop

Product Levels

ArcGIS for Desktop is available at different product levels, with increasing functionality.

- ArcReader (freeware, viewer) is a basic data viewer for maps and GIS data published in the proprietary Esri format using ArcGIS Publisher. The software also provides some basic tools for map viewing, printing and querying of spatial data. ArcReader is included with any of the ArcGIS suite of products, and is also available for free to download. ArcReader only works with preauthored published map files, created with ArcGIS Publisher.
- ArcGIS Desktop Basic, formerly known as ArcView, is the entry level of ArcGIS licensing offered. With ArcView, one is able to view and edit GIS data held in flat files, or view data stored in a relational database management system by accessing it through ArcSDE.
- ArcGIS Desktop Standard, formerly known as ArcEditor, is the midlevel software suite designed for advanced editing of spatial data published in the proprietary Esri format. It provides tools for the creation of map and spatial data used in GIS, including the ability of editing geodatabase files and data, multiuser geodatabase editing, versioning, raster data editing and vectorization, advanced vector data editing, managing coverages, coordinate geometry (COGO), and editing geometric networks. ArcEditor is not intended for advanced spatial analysis.
- ArcGIS Desktop Advanced, formerly known as ArcInfo, allows users the most flexibility and control in “all aspects of data building, modeling, analysis, and map display.” ArcInfo

includes increased capability in the areas of spatial analysis, geoprocessing, data management, and others.

Other desktop GIS software include ArcGIS Explorer and ArcGIS Engine. ArcGIS Explorer is a GIS viewer which can work as a client for ArcGIS Server, ArcIMS, ArcWeb Services and Web Map Service (WMS).

- ArcGIS Online is a web application allowing sharing and search of geographic information, as well as content published by Esri, ArcGIS users, and other authoritative data providers. It allows users to create and join groups, and control access to items shared publicly or within groups.
- ArcGIS Web Mapping APIs are APIs for several languages, allowing users to build and deploy applications that include GIS functionality and Web services from ArcGIS Online and ArcGIS Server. Adobe Flex, JavaScript and Microsoft Silverlight are supported for applications that can be embedded in web pages or launched as stand-alone Web applications. Flex, Adobe Air and Windows Presentation Foundation (WPF) are supported for desktop applications.

Components

ArcGIS for Desktop consists of several integrated applications, including ArcMap, ArcCatalog, ArcToolbox, ArcScene, ArcGlobe, and ArcGIS Pro. ArcCatalog is the data management application, used to browse datasets and files on one's computer, database, or other sources. In addition to showing what data is available, ArcCatalog also allows users to preview the data on a map. ArcCatalog also provides the ability to view and manage metadata for spatial datasets. ArcMap is the application used to view, edit and query geospatial data, and create maps. The ArcMap interface has two main sections, including a table of contents on the left and the data frame(s) which display the map. Items in the table of contents correspond with layers on the map. ArcToolbox contains geoprocessing, data conversion, and analysis tools, along with much of the functionality in ArcInfo. It is also possible to use batch processing with ArcToolbox, for frequently repeated tasks. ArcScene is an application which allows the user to view their GIS data in 3-D and is available with the 3D Analyst License. In the layer properties of ArcScene there is an Extrusion function which allows the user to exaggerate features three dimension-ally. ArcGlobe is another one of ArcGIS's 3D visualization applications available with the 3D Analyst License. ArcGlobe is a 3D visualization application that allows you to view large amounts of GIS data on a globe surface. The ArcGIS Pro application was added to ArcGIS for Desktop in 2015 February. It had the combined capabilities of the other integrated applications and was built as a fully 64-bit software application. ArcGIS Pro has ArcPy Python scripting for database programming.

Extensions

There are a number of software extensions that can be added to ArcGIS for Desktop that provide added functionality, including 3D Analyst, Spatial Analyst, Network Analyst, Survey Analyst, Tracking Analyst, and Geostatistical Analyst. Advanced map labeling is available with the Maplex extension, as an add-on to ArcView and ArcEditor and is bundled with ArcInfo. Numerous extensions have also been developed by third parties, such as the MapSpeller spell-checker, ST-Links PgMap XTools and MAP2PDF for creating georeferenced pdfs (GeoPDF), ERDAS' Image Analysis and Stereo Analyst for ArcGIS, and ISM's PurVIEW, which converts Arc- desktops into precise stereo-viewing windows to work with geo-referenced stereoscopic image models for accurate geo-database-direct editing or feature digitizing.

Address Locator

An address locator is a dataset in ArcGIS that stores the address attributes, associated indexes, and rules that define the process for translating nonspatial descriptions of places, such as street addresses, into spatial data that can be displayed as features on a map. An address locator contains a snapshot of the reference data used for geocoding, and parameters for standardizing addresses, searching for match locations, and creating output. Address locator files have a. loc file extension. In ArcGIS 8.3 and previous versions, an address locator was called a geocoding service.

Other Products

ArcGIS Mobile and ArcPad are products designed for mobile devices. ArcGIS Mobile is a software development kit for developers to use to create applications for mobile devices, such as smartphones or tablet PCs. If connected to the Internet, mobile applications can connect to ArcGIS Server to access or update data. ArcGIS Mobile is only available at the Enterprise level.

Server GIS products include ArcIMS (web mapping server), ArcGIS Server and ArcGIS Image Server. As with ArcGIS Desktop, ArcGIS Server is available at different product levels, including Basic, Standard, and Advanced Editions. ArcGIS Server comes with SQL Server Express DBMS embedded and can work with enterprise DBMS such as SQL Server Enterprise and Oracle. The Esri Developer Network (EDN) includes ArcObjects and other tools for building custom software applications, and ArcGIS Engine provides a programming interface for developers.

For non-commercial purposes, Esri offers a home use program with an annual license fee.

ArcGIS Engine

The ArcGIS Engine is an ArcGIS software engine, a developer product for creating custom GIS desktop applications.

ArcGIS Engine provides application programming interfaces (APIs) for COM,. NET, Java, and C++ for the Windows, Linux, and Solaris platforms. The APIs include documentation and a series of high-level visual components to ease building ArcGIS applications.

ArcGIS Engine includes the core set of components, ArcObjects, from which ArcGIS Desktop products are built. With ArcGIS Engine one can build stand-alone applications or extend existing applications for both GIS and non-GIS users. The ArcGIS Engine distribution additionally includes utilities, samples, and documentation.

One ArcGIS Engine Runtime or ArcGIS Desktop license per computer is necessary.

Criticisms

Esri's transition to the ArcGIS platform, starting with the 1999 release of ArcGIS 8.0, rendered incompatible an extensive range of user-developed and third-party add-on software and scripts. A minority user base resists migrating to ArcGIS because of changes in scripting capability, functionality, operating system (Esri developed ArcGIS Desktop software exclusively for the Microsoft Windows operating system), as well as the significantly larger system resources required by the ArcGIS software. Esri has continued support for these users. ArcView 3.x is still available for purchase, and ArcInfo Workstation is still included in a full ArcGIS ArcInfo licence to provide some editing and file conversion functionality that has not been included to date in ArcGIS. Other issues with ArcGIS include perceived high prices for the products, proprietary formats, and difficulties of porting data between Esri and other GIS software.

ArcGIS Model Builder

Model Builder in ArcGIS automates your GIS workflows. A model usually consists of at least three elements:

- Input Data (blue squares),
- Geoprocessing Tools (yellow circles),
- Output Data (green squares).

In our model builder example, here are the two inputs:

Input 1. Pipeline alignment (red and black-dashed line)

Input 2. Endangered bird nests

And here's how these inputs look like on a map:

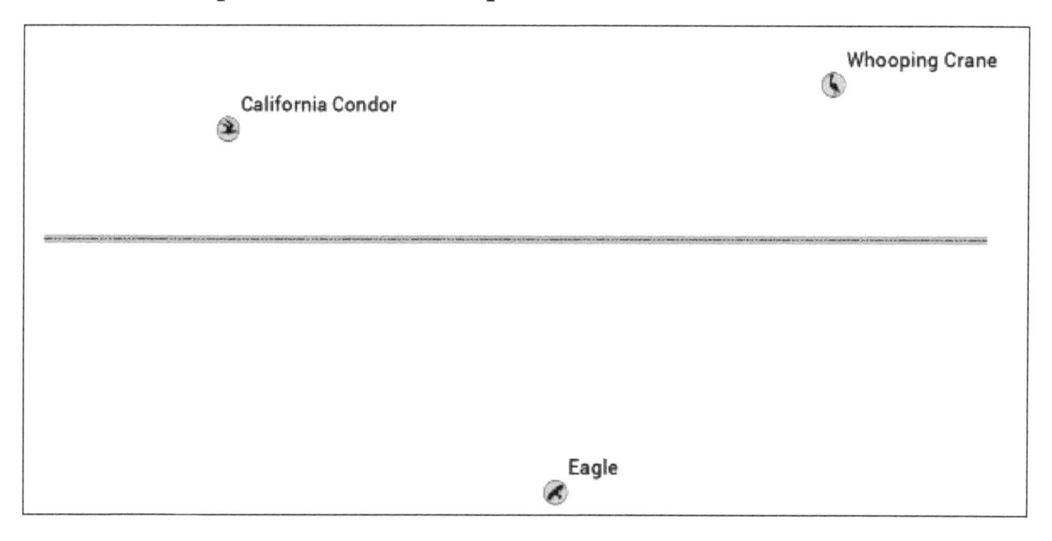

Tool 1. Buffer pipeline 500 meters

Tool 2. Intersect bird nests with pipeline buffer

Tool 3. Generate report as spreadsheet

Output 1. Buffer, Bird Nests in Buffer, Report

Step 1 Open Model Builder

First, click the Model Builder icon to start creating a model. This will open the Model Builder window for you to start adding your geoprocessing tools.

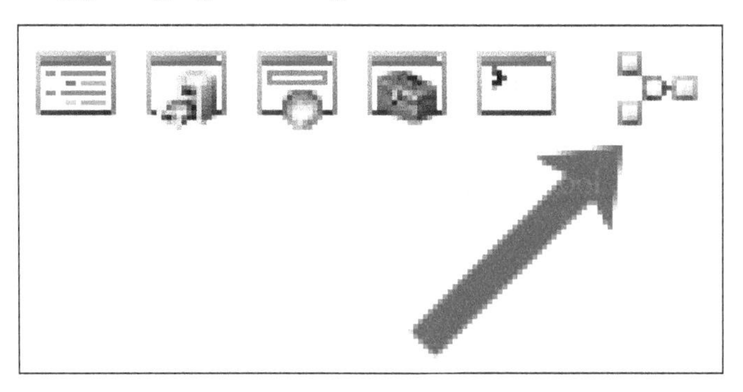

Step 2 Add tools to Model Builder

Drag and drop the buffer, intersect tool and Excel spreadsheet conversion tools from ArcToolbox.

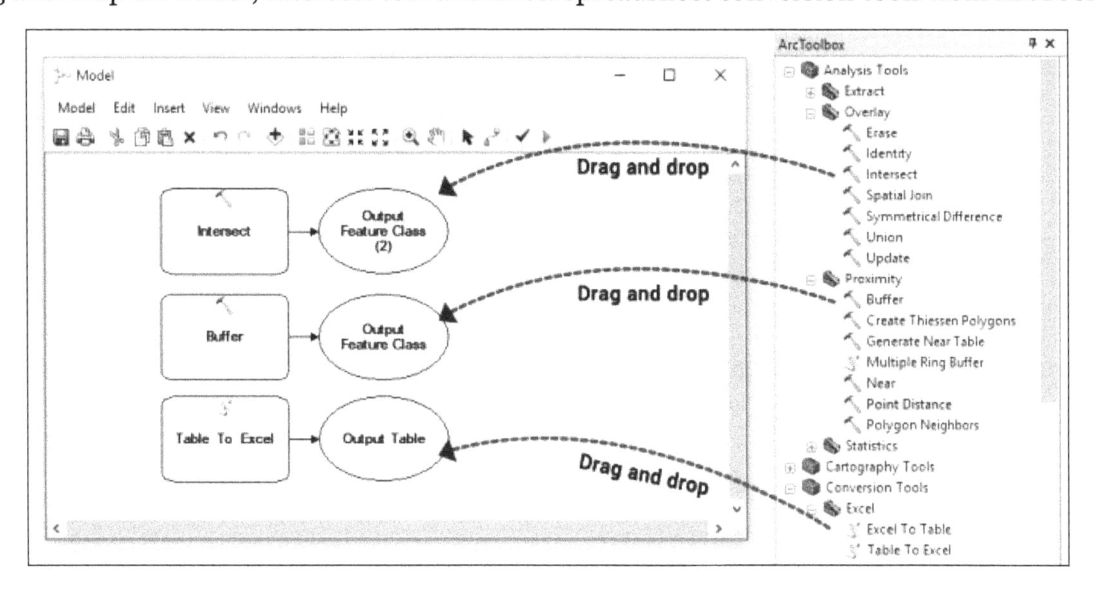

Step 3 Enter tool parameters

Double-click each tool (rectangle) and enter in the following inputs and outputs:

Buffer

Input Features: Pipeline Realignment

Output Feature Class: C:\Temp\Pipeline_Buffer.shp (Make sure this directory exists or create a new folder if necessary)

Linear Distance: 500 meters

Intersect:

Input Features: Bird Nest and Pipeline_Buffer.shp

Output Feature Class: C:\Temp\BirdNest_Intersect.shp

Table To Excel:

Input Table: BirdNest_Intersect.shp

Output Excel File: C:\Temp\Report.xls

After entering the inputs and outputs for each tool, the Model Builder window will look like the screenshot below. However, it looks disorganized and the auto-layout tool can help clean everything up.

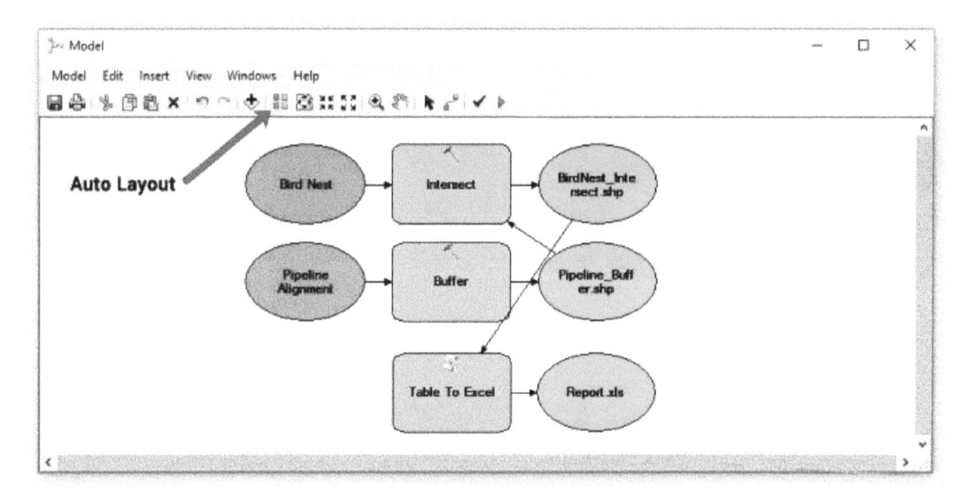

Step 4 Auto layout

After you click the auto-layout button, the geoprocessing tools, inputs and outputs will rearrange themselves neatly in the model builder window.

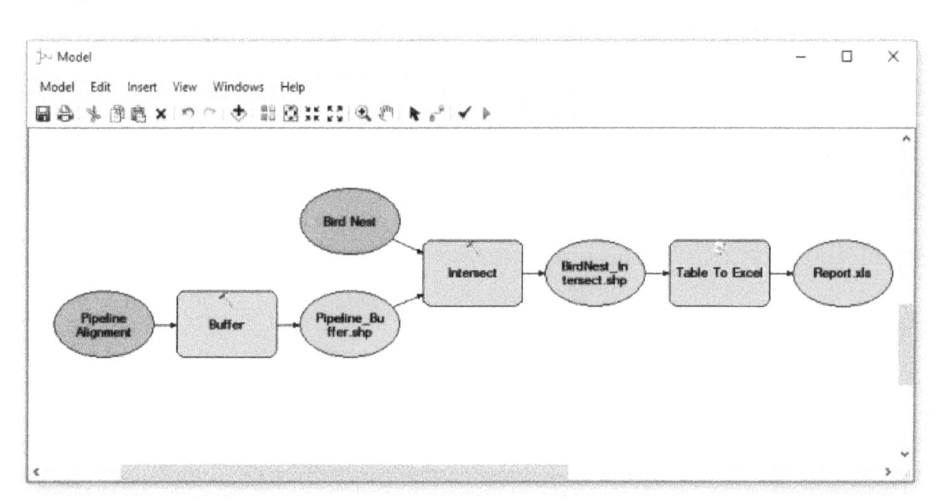

Model Builder Example: Running the Model

Of course, we have only built the model up to this point. In order to create the report, we still have to run the model. You can run the model by clicking the 'Play' button.

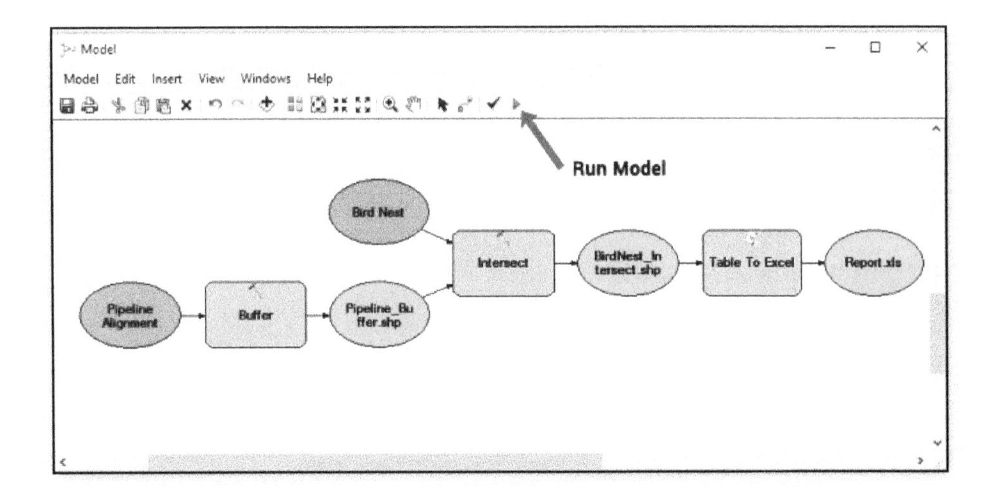

After you run your model, you can add your output feature classes to ArcMap to verify it is correct. By double-clicking the report.xls, it lists all bird nests within 500-meters of the pipeline. In this case, it's only the California Condor.

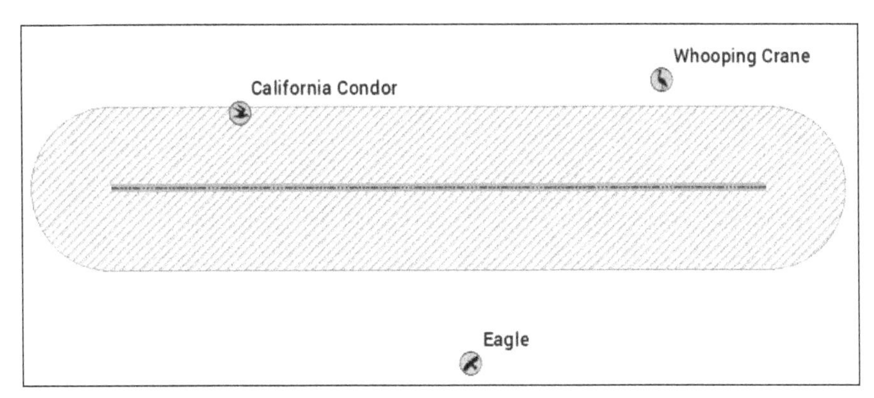

Another thing to note is that after you run your model, circles and squares in the model will have shadows. Check (C:\Temp\) to see each product it creates (buffer, intersect and report).

But really, the only output you need is just the report. The buffer and intersect are just temporary (intermediate) data sets. After running a model from the ModelBuilder window, it doesn't delete all the intermediate outputs (buffer and intersect) that it creates.

So if you want to re-run the model, you can manually go into this folder ("C:\Temp\") and delete these feature classes. Also, you may have to click the "checkmark" icon next to the "auto-layout" icon to validate the model.

Model Builder Toolbox: Setting up Parameters/Variables

Let's say another group in your company wants to use this model for the same purpose. But their input data and output data have completely different names.

Instead of static input and output, you need them to be dynamic as parameters. When you set the input data or output data as parameters, users can enter their own data and set their own output paths.

How do you set up model builder parameters?

Step 1 Set model parameters for inputs

Right-click your two input data sets (bird nests and pipeline) and click 'Model Parameter'

Step 2 Set model parameter for output

Right-click your output data set (report) and click 'Model Parameter'

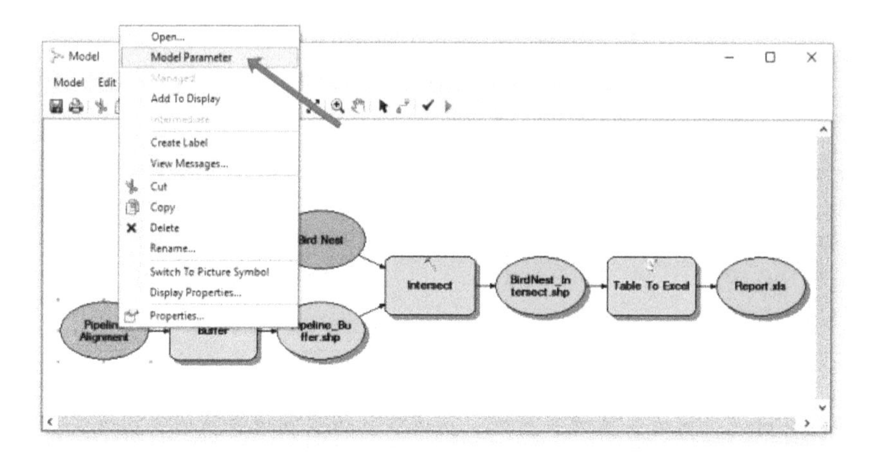

But there's one more thing that we'll have to change – how it handles intermediate data. ArcGIS Model Builder has a handy command called 'in_memory' to handle it.

For each temporary output (buffer and intersect), we can turn these into temporary datasets with 'in_memory'. For example, double-click the green box (pipeline_buffer) and change it as follows:

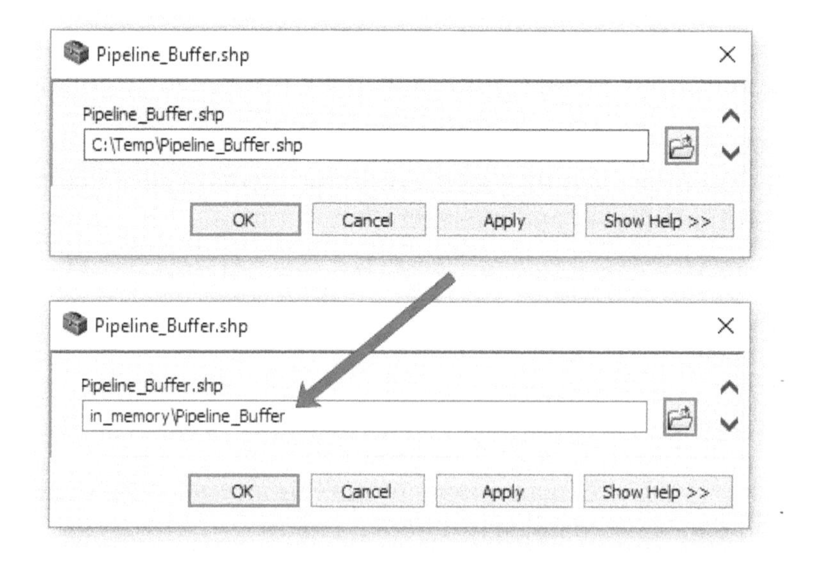

After this step, your model should have "P" next to each model parameter. Also, your temporary buffer and intersect output should have 'in_memory' for the output.

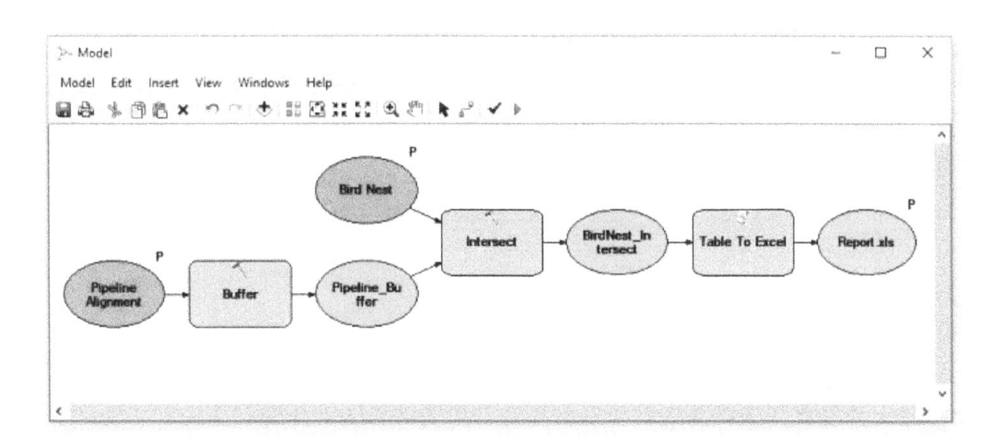

Delete all the files in (C:\Temp\). Now, when you click 'Run Model', it will only create the XLS report. And this is because you set the buffer and intersect as temporary data (in_memory).

How to Create a Toolbox in ArcGIS

Now that we've set the input and output as parameters, wouldn't it be nice if this could be a toolbox on its own? Toolboxes are convenient to share. They look professional. And they are easy to create.

First, let's empty the parameters so that when we share it, there isn't a default path. Double-click each model parameter in model builder and empty it.

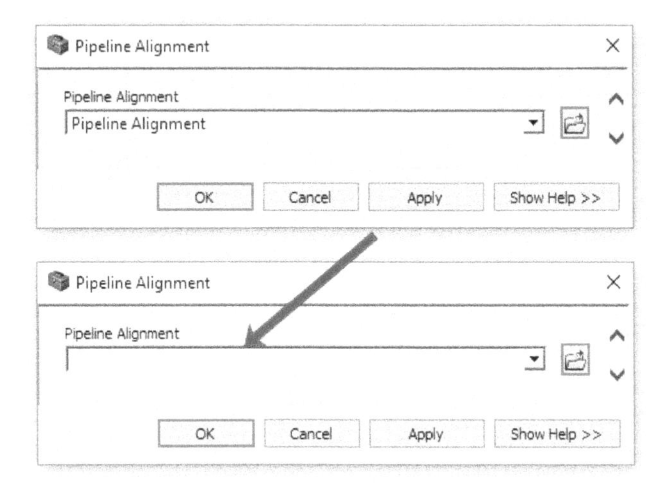

Once you've emptied each parameter, the model should look white like in the screenshot below:

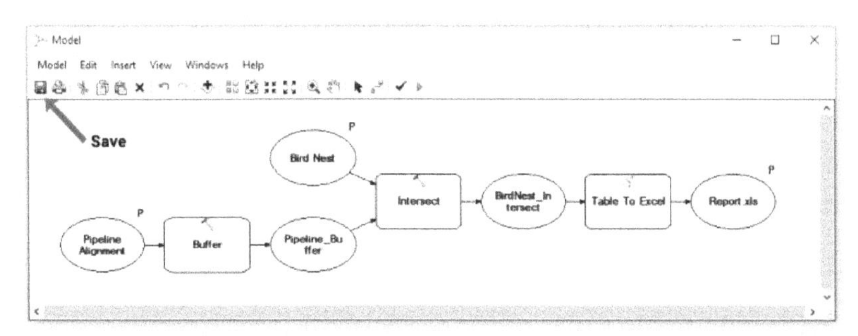

Now, you'll want to save your model. But you're going to need a toolbox to do so. After clicking "Save", click the new toolbox icon in the top-right of the dialog box.

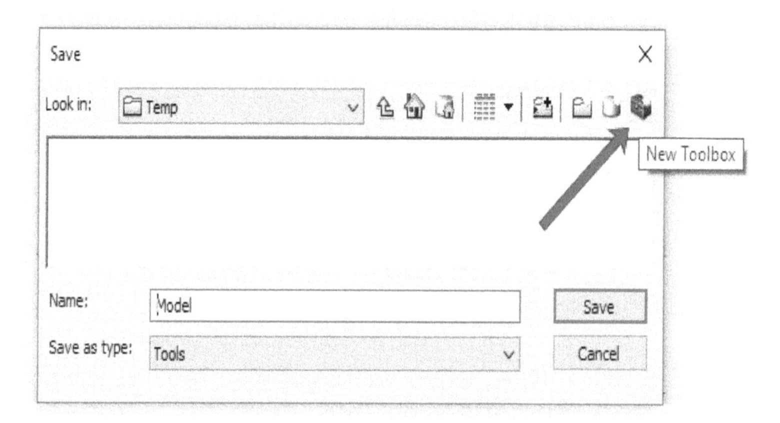

You can name the new toolbox "Pipeline.tbx". Now, double-click this new toolbox and save the model within the toolbox.

In ArcCatalog, locate where you just saved this toolbox. Expand the toolbox so you can see your new toolset.

Double-click the new toolset that you have created. Now, that we have created parameters and emptied them, you can now dynamically enter in your inputs and outputs.

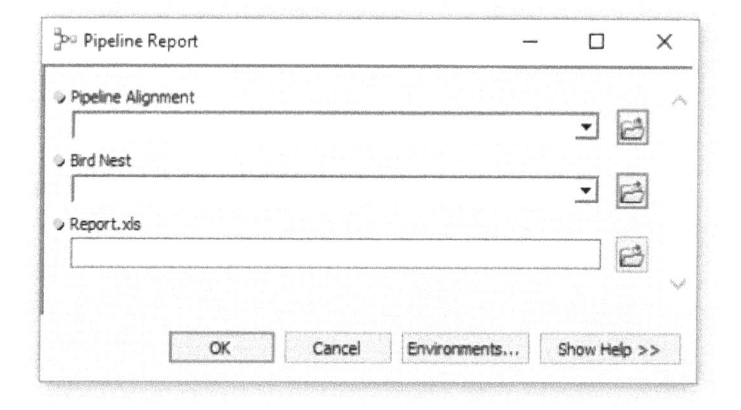

COMMERCIAL GIS SOFTWARE

Businesses and GIS users around the world are embracing commercial GIS software because:

- Customers prefer 24/7 business-class support – and someone to yell at when it's not working properly.
- The software and GIS file formats have become the de facto standard in the industry.
- Less training is required because universities select it for teaching.

ArcGIS (Esri)

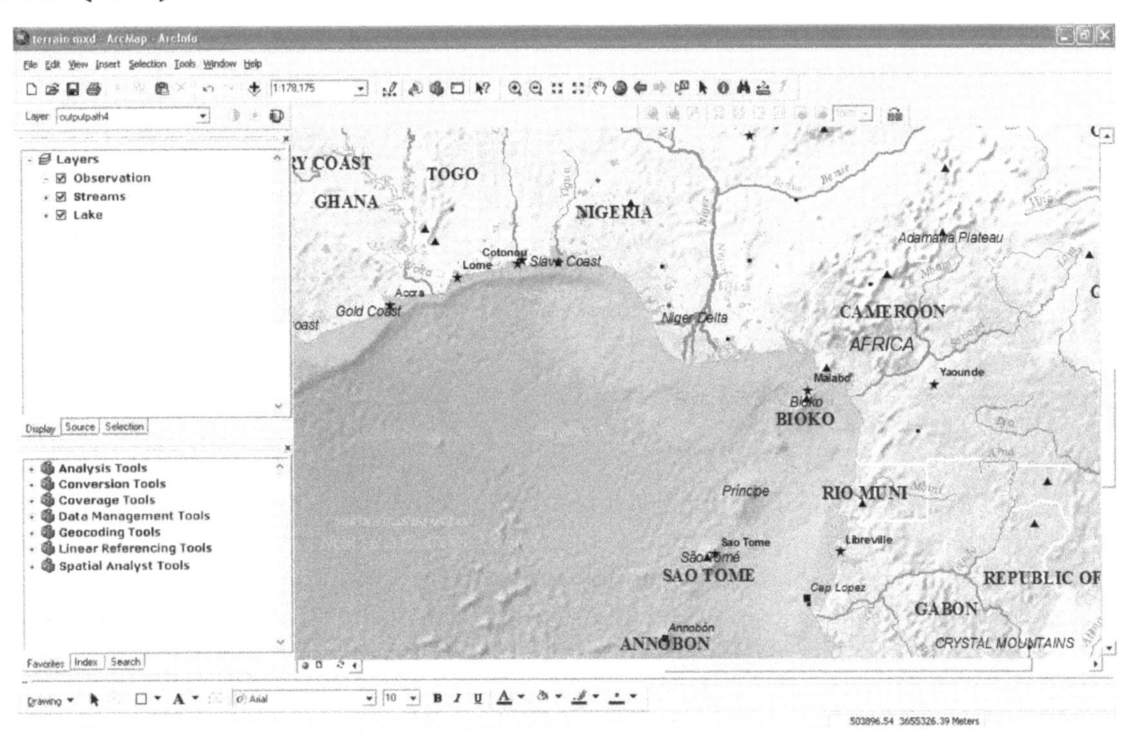

ArcGIS 9.x.

Esri busted onto the geospatial scene in the 1970s. They've pulled ahead as being the market leader in commercial GIS software and hasn't looked back since. The fine-tuned Esri ArcGIS is the most innovative, cutting-edge GIS software in the industry. When you look at the amazing features in the whole ArcGIS suite, it begins to make sense why they are so good at what they do. ArcGlobe and ArcScene bring objects to life in a separate 3D interface. Furthermore, ArcEarth has an interface similar to Google Earth that is part of the Esri family of software. Standard mapping is top-notch in ArcMap. They've perfected automatic map production with data-driven pages, ultimately saving time and money. The extensions are amazing. Network analyst is one of a kind, and nothing comes close from other commercial GIS software.

ArcGIS simplifies geostatistics by actually teaching you statistics giving the user more informed decisions. Scripting and model builder deserves all the rave it gets. Stringing together models is an easy process after a bit of practice. Esri GIS software provides scalability, interoperability, stability, flexibility with their cartographic, modelling and spatial features. There are so many reasons why Esri should be the choice for your organization.

Geomedia (Hexagon Geospatial)

Geomedia has been the main rival to ArcGIS for years, even decades. Previously owned by Intergraph, now Hexagon Geospatial Geomedia has a 40 year history in the GIS industry. Government, infrastructure and military groups are embracing Geomedia because of its solution-driven approach. Geomedia is a comprehensive commercial GIS software package because how it provides advanced data management, visualization, analysis and cartographic tools. It's a powerful and flexible solution to extract actionable information.

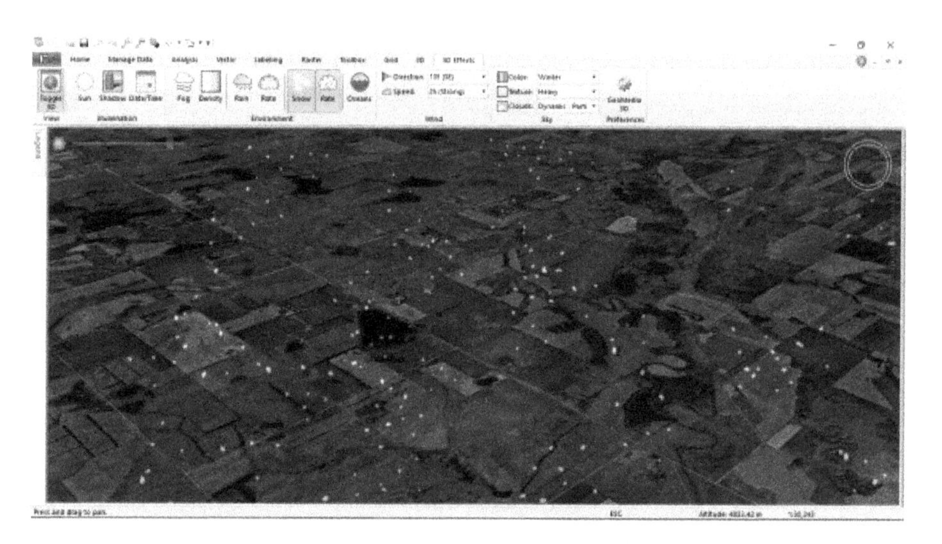

The quickness of Geomedia 3D displaying three-dimensional models will startle you. The rendering is accurate, powerful and beautiful. Geomedia 3D delivers. It has smart data capture and validation tools because data integrity is important to all organizations. Geomedia has cadastral data management that ensures completeness and robust commands. For example, it has tools for conflation (selectively combine data with conflation rules to increase the overall quality), 3D point cloud (such as LiDAR) and mobile capabilities with MapWorks. But this is just scratching the surface of tools and options available.

MapInfo Professional (Pitney Bowes)

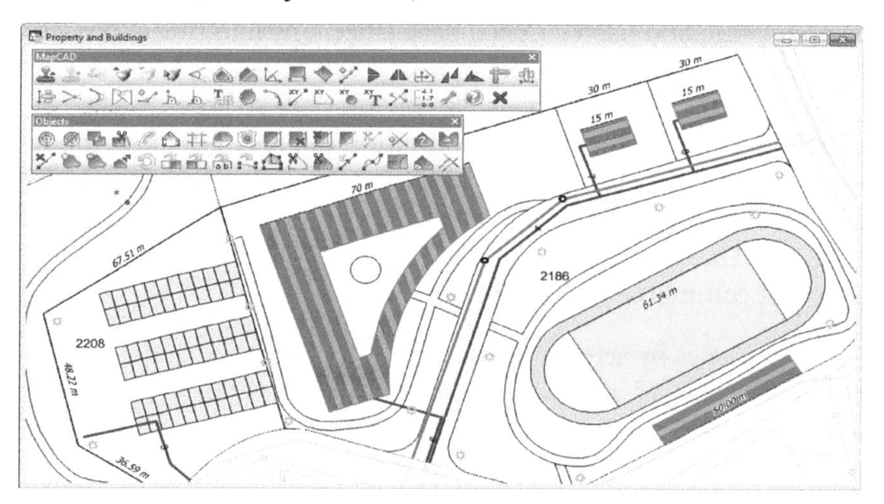

MapInfo Professional has come a long way in terms of proprietary GIS software. It has a clear focus on location intelligence. Locate optimal positions for retail stores, uncover geographic patterns and improve insurance risk with GIS insurance mapping. Usability means productivity – This is a core philosophy for MapInfo. It comes naturally as naturally to you as breathing. Ultimately, this helps drive greater profits.

The latest version brings improvements in cartographic output. It's smarter with labeling, legend and scale bars. It brings 64-bit compatibility, enhanced layouts and much more. It's a full-blown GIS software package always mapping out new ways to innovate.

Global Mapper (Blue Marble)

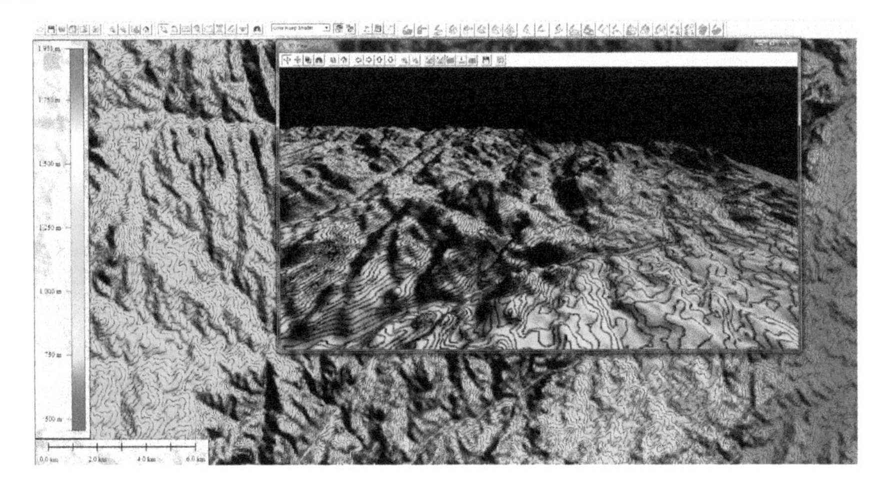

Global Mapper was originally developed by the USGS. It has evolved into a GIS commercial software product competing with the GIS industry giants. If there was a 'Swiss-army knife' for GIS software, Global Mapper would be it. This commercial GIS software is flexible enough to satisfy both the beginner and advanced GIS user. Its core applications reach back to its roots of working with elevation data. These specialized tools include 3D rendering, watershed delineation and LiDAR handling. Global Mapper offers decent interoperability, map publishing tools in a flexible map viewer interface. It reads a large number of geospatial formats. Print layouts and symbolization variety are some of the only complaints of Global Mapper. Overall, it's an affordable commercial GIS software – at a fraction of the cost of other GIS proprietary software.

Manifold GIS (Manifold)

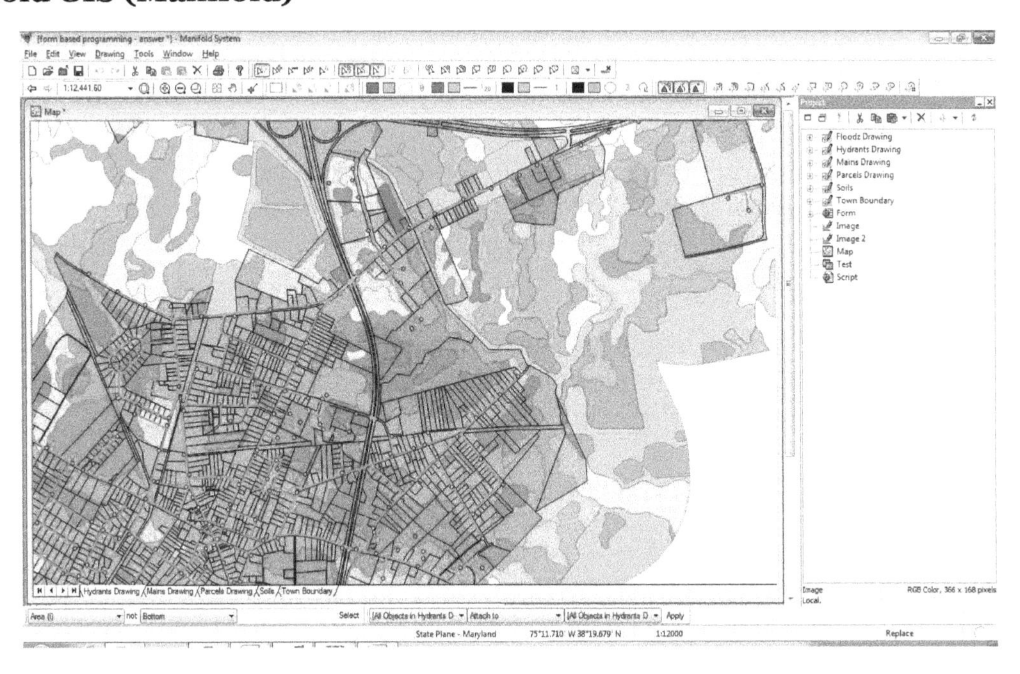

How awesome would it be to get all the features in a full-blown GIS package but pay half the price? Manifold GIS is all about delivering world-class GIS commercial software. It's intuitive interface enables GIS users to seamlessly work endlessly in a 64-bit GIS environment. Manifold GIS is a combination of mapping, CAD, DBMS and image processing. It's stable with a wide range of functions. Although limited on cartography options, you can still deliver some spectacular map products. The 3D bar charts in Manifold GIS are drool-worthy. GIS bargain hunters like its unbeatable price/value ratio. Manifold GIS may be the only GIS software you need.

Smallworld (General Electric)

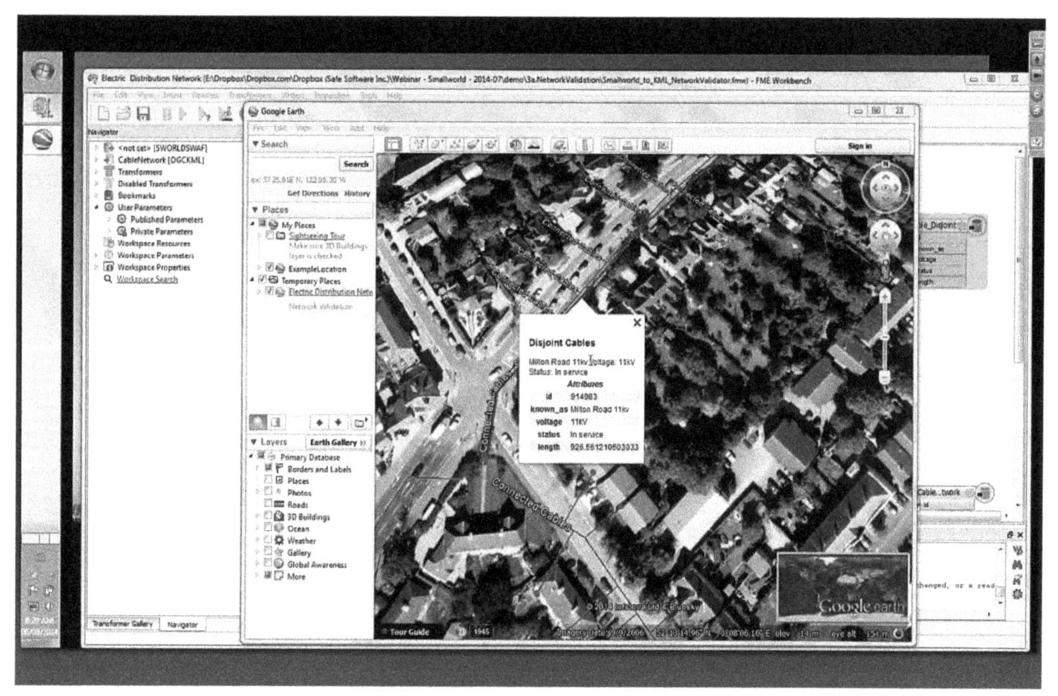

GE SmallWorld

GE Smallworld makes all the difference for tech-savvy GIS users working in utilities. This GIS commercial software lets you better manage, design and plan any type of network infrastructure with its centerpiece Smallworld Core. Smallworld Core gives you the capability to capture and visualize complex spatial networks. It provides the foundation to manage the life-cycle of network assets. From network planning, design, build, operations and maintenance, SmallWorld holds your hand every step of the way.

This commercial GIS software is highly expandable and is widely used by electrical, telecommunication, gas, water utilities. Outside this realm, you may want to look elsewhere. Some of its key strengths are its versioning, available architecture and expandability.

Bentley Map

We step into a world of CAD and GIS fusion. In the Bentley world, their 2D and 3D viewing provides a foundation for top-notch visualization. Fly-throughs in Bentley Map are a heaven-sent opportunity. Shadow studies, 3D intersects and clash detection come with ease with Bentley's 3D tools. With Bentley Map, you can access, share, edit and analyze CAD, BIM and GIS data. It advances GIS for infrastructure from basic mapping to tracking infrastructure life-cycle. You can do away with the divider from GIS and engineering through Bentley Map. Connect to other GIS proprietary software formats such as Esri file geodatabases, MapInfo and other CAD engineering company formats. It's an optimal choice for organizations heavy on the CAD end looking to expand into GIS.

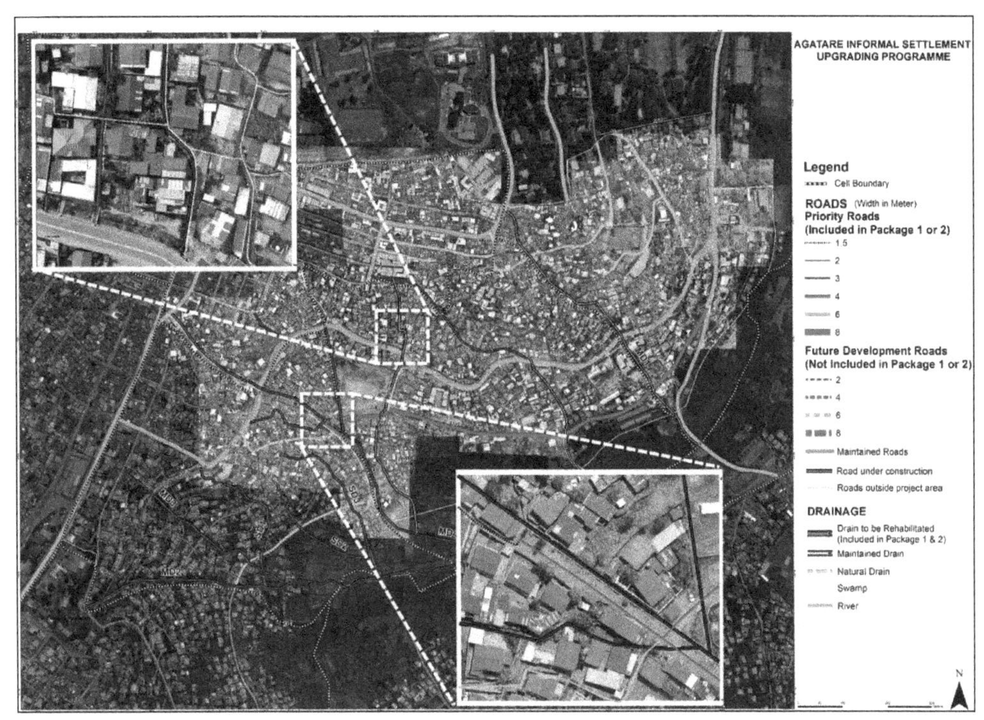

Bentley Map.

MapViewer and Surfer (Golden Software)

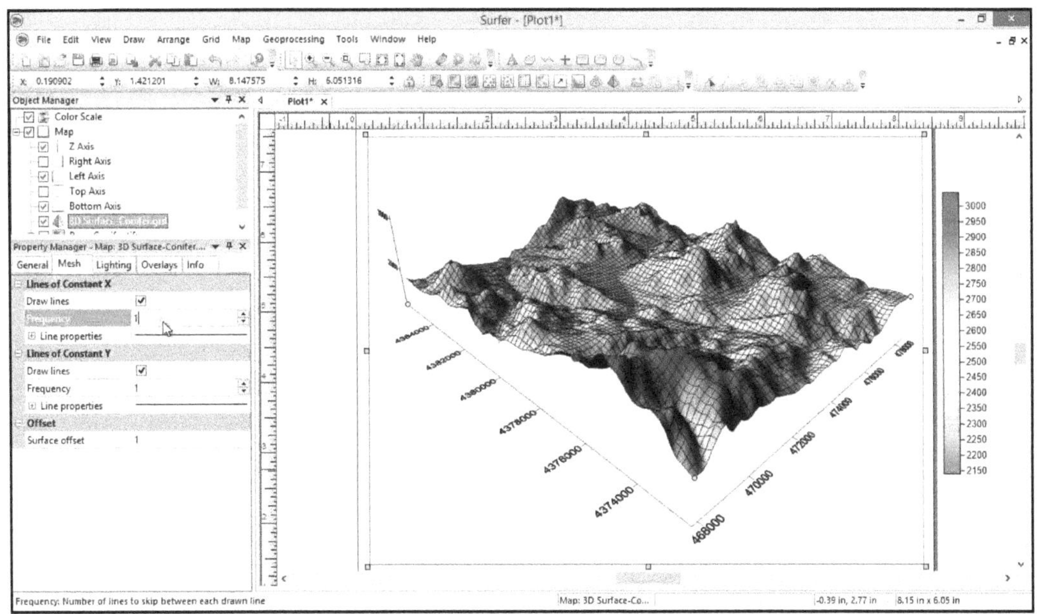

MapViewer.

Meet the Golden Software family of GIS commercial software for your geospatial needs. Surfer is a powerful contouring, gridding and surface making package. Its versatility lets you create some unique maps in XYZ directions. There's a short learning curve for this commercial GIS software.

The heart of Surfer is visualizing 3D data. Surfer delivers a total of 11 map type options including wireframe, 3D surface maps and vector-scale maps. MapViewer lets you publish and share professional maps. This includes pin, contour, symbol, density, hatch, prism, territory and line graph maps. Geocoding is available in MapViewer. But it's not a forte. Overall, mapping can be difficult at times but with the Golden Software family it doesn't have to be.

Maptitude (Caliper Corporation)

Map with confidence using Maptitude. You can produce quick and beautiful maps and tell the story that your spreadsheets can't. Maptitude is a way to find geographic patterns and trends with the vast amounts of spatial data you work with. Hidden in the numbers, Maptitude unlocks how geography can affect businesses. Although less on GIS processing tools, it has some tools to get the job done like determining drive-time and identifying hot spots. Maptitude has some neat mapping outputs like 3D prism maps, scaled-symbol and bar chart maps. It comes packed with population, household, employment demographics and education data. It's easy to get into and fun to chart out your geographic data with Maptitude.

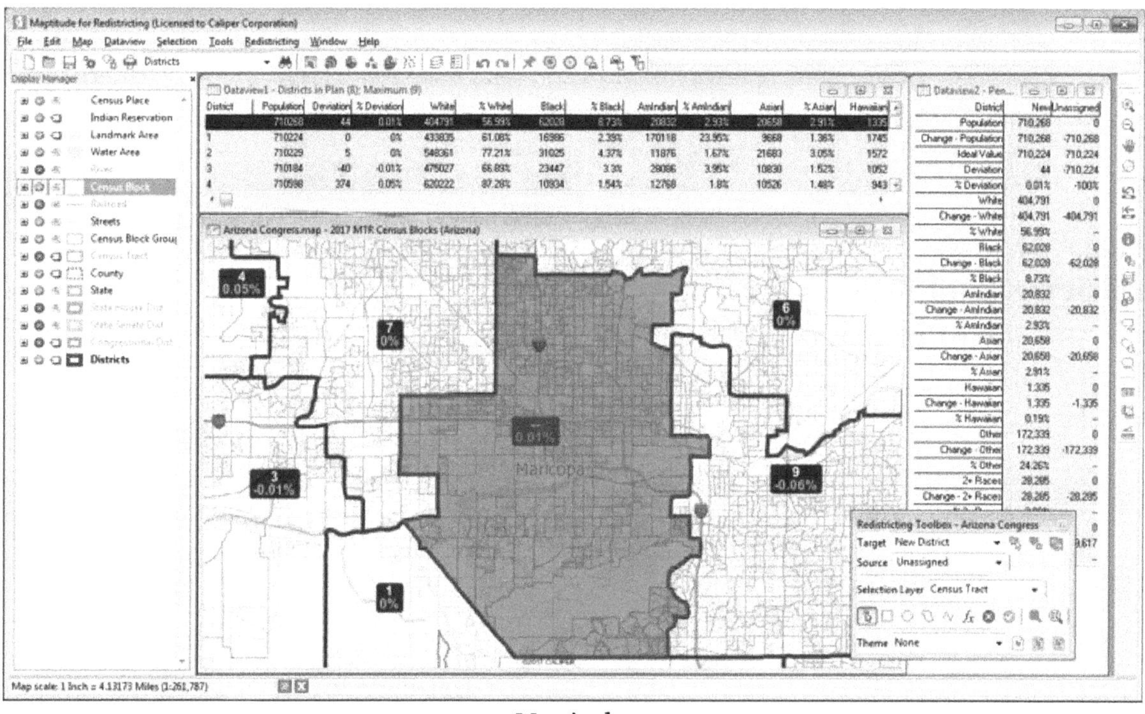

Maptitude

SuperGIS (Supergeo Technologies)

The look and feel of SuperGIS eerily looks similar to ArcGIS. The data/layout view, the on-the-fly coordinate systems, the extension options. SuperGIS was fast at processing geospatial data. It buffered, clipped and merged faster than some of the other commercial GIS software giants on this list. And when you're on the clock working in the GIS industry, time is a precious resource. Tools outclocked ArcGIS at times. SuperGIS has some unique features like the biodiversity analysis. There are analyses for patches and biodiversity classes on the landscape. It's always fascinating seeing the 3D viewers available. SuperGIS 3D Earth Server enables GIS users to get a bird's-eye view of a city or even manage underground utilities.SuperGIS is a cheap alternative. It's fast. And it has a whole boatload of GIS functionality.

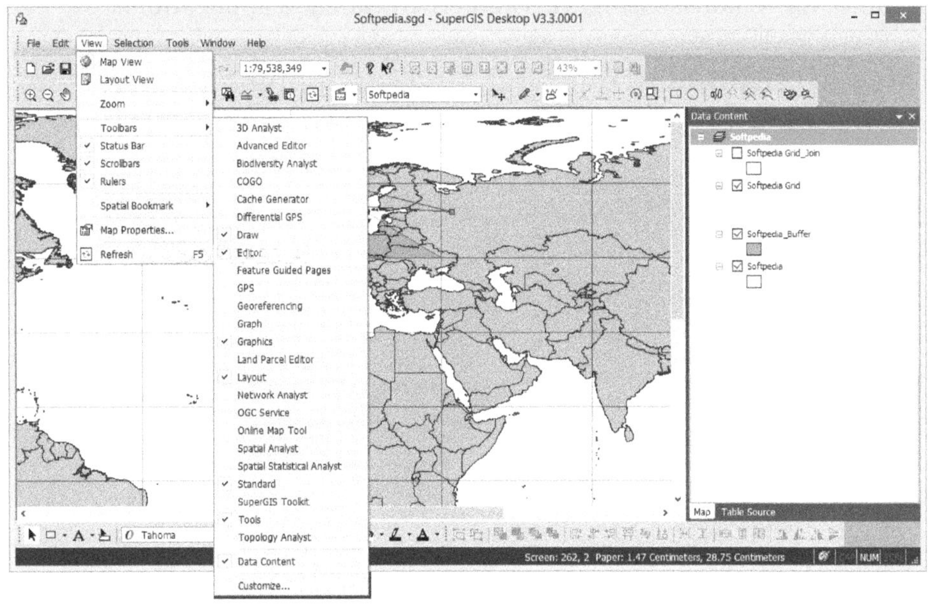

SuperGIS

IDRISI (Clark Laboratories)

Clark Laboratories IDRISI more specializes in remote sensing than anything else. It has a strong foothold in remote sensing and commercial GIS software. It's used in colleges and universities from the get-go.

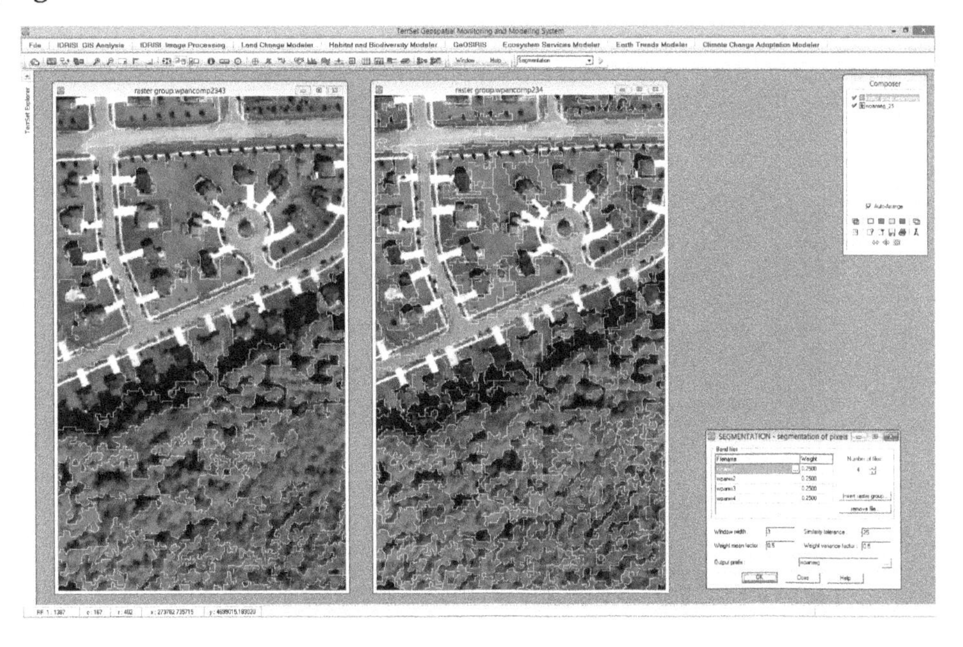

IDRISI boasts a 300-set of GIS processing tools. Although primarily oriented towards raster processing, IDRISI has adapted into a commercial GIS software suite. What's really neat about IDRISI is the add-ons. The Land Change Modeler add-on models trends on the landscape for decision support. The Habitat and Biodiversity Modeler shows patterns for habitat assessment. There are add-ons for climate change, ecosystem and Earth trends. This goes to show how IDRISI software is the perfect choice for remote sensing, GIS and environmental applications.

AutoCAD Map 3D (Autodesk)

AutoCAD isn't just about drafting. AutoCAD Map 3D bridges the gap between CAD and GIS. For those already using Autodesk, this commercial GIS software gives a familiar interface to manage spatial data and underlying asset information. With its growing range of tools to choose from, you can also produce some beautiful cartographic output. As the name suggests, AutoCAD Map 3D seamlessly renders spatial data in three dimensions. Whether you're working in oil and gas, public utilities or the mining industry, 3D mapping is an advantageous feature. AutoCAD certainly delivers in that respect and much more.

Tatuk GIS

TatukGIS

TatukGIS comes packaged as different versions. The slimmed down TatukGIS Viewer is completely free for everyone to use. Their commercial GIS software TatukGIS Editor gives a great deal more features including scripting/customization, 3D mapping and advanced data editing. With state-of-the-art editing functionality, one of the bright spots for TatukGIS is the topology and error-checking tools. It has on-the-fly projections with support of over 900 datums. It supports most database engine for a wide open choice for enterprise-level data storage. The name TatukGIS comes from Tatuk Lake in British Columbia, Canada. And this is a brand you should remember for proprietary GIS software.

MicroImages (TNTgis)

TNTmips, TNTedit, TNTview and TNTscript is the MicroImages commercial GIS software family. TNTview is the bare bones version from MicroImages. But there's nothing bare about it. It comes with some fairly advanced options like stereo and 3D viewing. TNTview can be used for map design and thematic cartography. One cut above is TNTeditor. Take all the features from TNTview and add data editing. Create, georeference and edit imagery with this powerful GIS commercial software. TNTmips is th complete proprietary GIS software solution. It has LiDAR support, terrain analysis, web map publishing and a load more of useful GIS tools. Finally, TNTscript lets you string together sets of GIS processing tools. Process GIS data locally or through cloud computing resources. This gives automation a new level to complete your work flow.

MapMaker Pro (MapMaker)

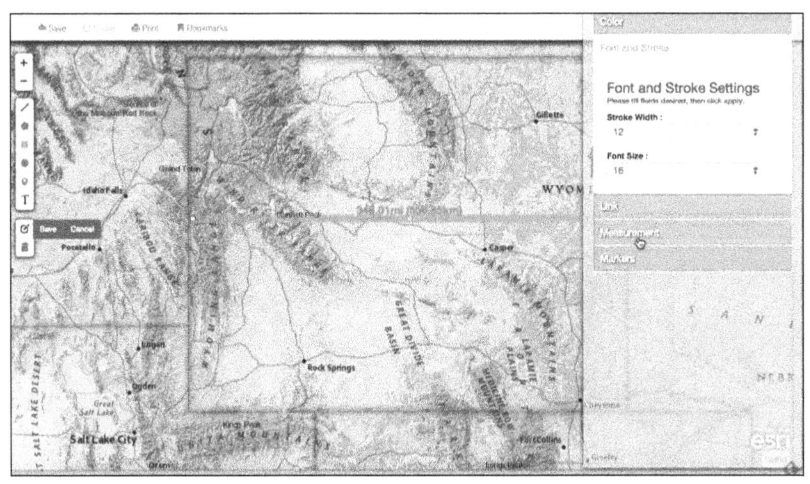

MapMaker Pro

We tested the software and realized instantly – the goal of MapMaker Pro is for anyone to be able to create a map with ease. We were instantly in love. It's a low-cost solution for those who need to put spatial data on a map. It accepts the most common GIS formats. There are capabilities to manipulate 3D, GPS, vector and raster data. With this commercial GIS software, you don't get a whole ton of options to manipulate and analyze GIS data. But for this looking for a quick and cheap solution to mapping out data, MapMaker Pro might be the answer you're looking for.

XMap (Delorme)

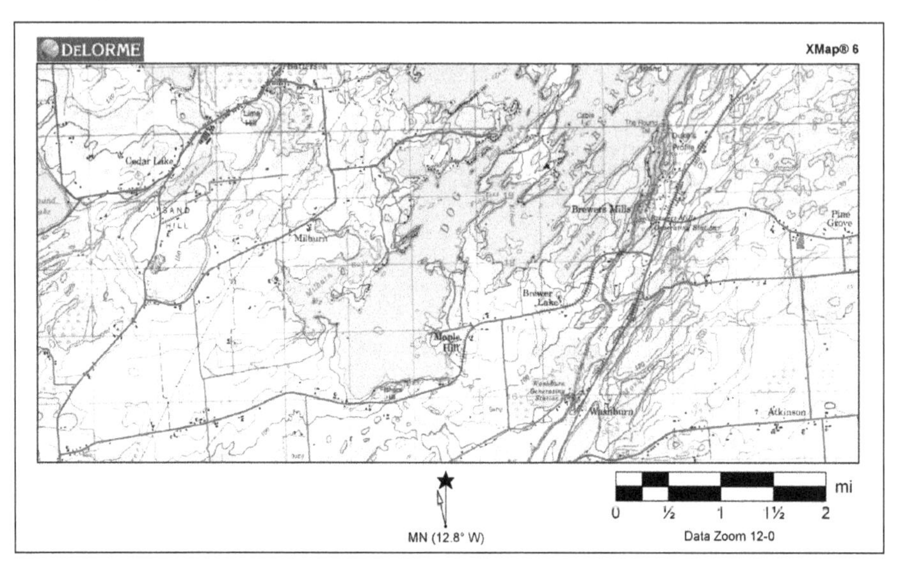

XMap is a user-friendly, low-cost, all-in-one GIS software. But it's mostly for simplifying data collection. XMap bridges the gap between data collection and field staff. For example, users can create forms to easily collect data in the field. Other key points is that XMap Professional is a GIS data viewing application. It can be fully integrated in an Esri environment. XMap GIS Editor offers tools to import, create, query and edit data for small-scale GIS operations. Although primarily used for field activities, XMap is a commercial GIS software option to also edit, manage and visualize geospatial data.

MapRite (Envitia)

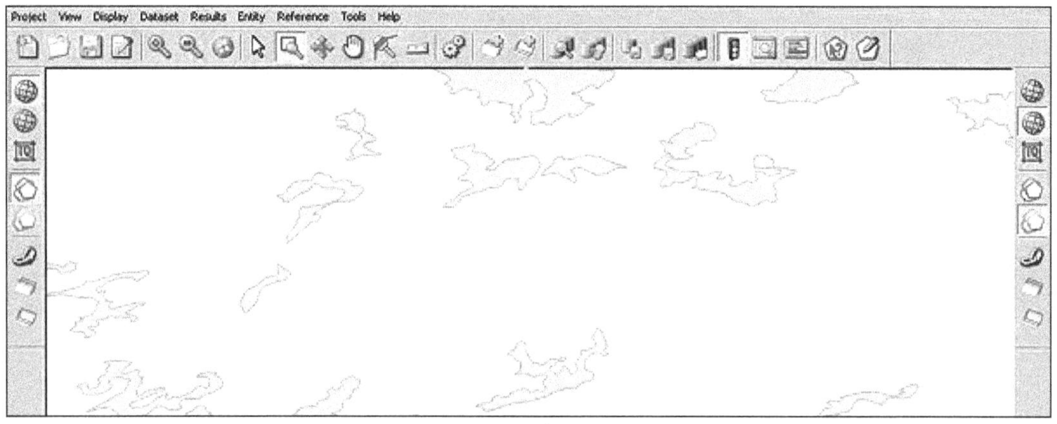

MapRite

Conflationers need to take a look at MapRite from Envita. Envita MapRite is a market leader in conflation and data cleansing. It seamlessly corrects location errors with a bit of user guidance. It boasts some powerful clientele. Customers like Land Registry, Scottish Power and Network Rail are using MapRite because of its ability to maintain accurate records. It has an automatic service for managing map change and reporting features. With superior annotation handling and a bunch of other features, MapRite puts the power of GIS in the hands of the user.

References

- Gis-data-capture: elegantsolutions.us, Retrieved 2 June, 2019
- What-is-a-geoportal-and-the-geoportal-server, geoportal-server: github.com, Retrieved 22 May, 2019
- Remote-sensing-earth-observation-guide: gisgeography.com, Retrieved 23 February, 2019
- Wade, tasha; sommer, shelly, eds. (2006). A to z gis: an illustrated dictionary of geographic information systems. Esri press. Isbn 9781589481404
- Santafe, meetings, gishydro, maidment, prof: utexas.edu, Retrieved , 10 July 2019
- Smith, susan (may 2004). “dr. David maguire on the arcgis 9.0 product family release”. Gis weekly. Retrieved 2013-04-14
- Deep-machine-learning-ml-artificial-intelligence-ai-gis: gisgeography.com, Retrieved 13 March, 2019
- Fee, James (2006-10-02). “Do you still use ArcView 3.x?”. Spatially Adjusted. Archived from the original on 23 February 2008. Retrieved 2008-02-05

4

Spatial Analysis in Geographic Information System

The set of formal techniques that study entities using their geometric, geographic and topological properties are referred to as spatial analysis. The chapter closely examines the key concepts related to spatial analysis such as spatial data and its types provide an extensive understanding of the subject.

SPATIAL DATA AND ITS TYPES

Spatial data is always being considered whether you know it or not. Spatial data, also known as geospatial data, is a term used to describe any data related to or containing information about a specific location on the Earth's surface.

Spatial data can exist in a variety of formats and contains more than just location specific information. To properly understand and learn more about spatial data, there are a few key terms that will help you become more fluent in the language of spatial data.

Vector

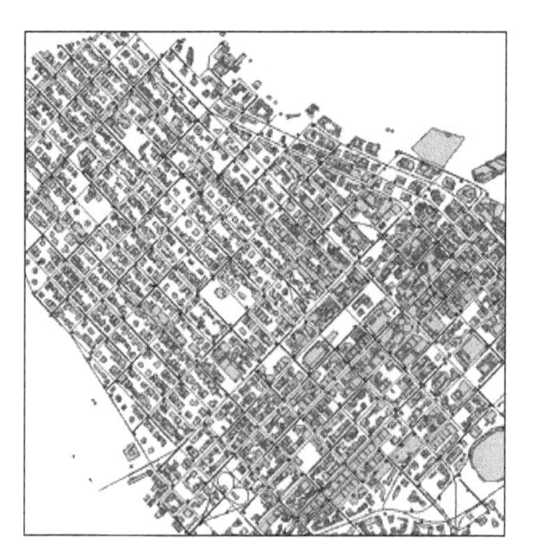

Vector data is best described as graphical representations of the real world. There are three main types of vector data: points, lines, and polygons. Connecting points create lines, and connecting lines that create an enclosed area create polygons. Vectors are best used to present generalizations of objects or features on the Earth's surface. Vector data and the file format known as shapefiles (.shp) are sometimes used interchangeably since vector data is most often stored in. shp files.

Raster

Raster data is data that is presented in a grid of pixels. Each pixel within a raster has a value, whether it be a colour or unit of measurement, to communicate information about the element in question. Rasters typically refer to imagery. However, in the spatial world, this may specifically refer to orthoimagery which are photos taken from satellites or other aerial devices. Raster data quality varies depending on resolution and your task at hand.

Attributes

Spatial data contains more information than just a location on the surface of the Earth. Any additional information, or non-spatial data, that describes a feature is referred to as an attribute. Spatial data can have any amount of additional attributes accompanying information about the location. For example, you might have a map displaying buildings within a city's downtown region. Each of the buildings, in addition to their location, may have additional attributes such as the type of use (housing, business, government, etc.), the year it was built, and how many stories it has.

Geographic Coordinate System

To identify exact locations on the surface of the Earth, a geographic coordinate system is used. Normally, an x and y-axis are used in mathematical systems, but in geography, the axes are referred to as lines of latitude (horizontal lines that run east-west) and longitude (vertical lines that run north-south). Each axis represents the angle at which that line is oriented with respect to the center of the Earth, and so the units are measured in degrees (°).

Georeferencing and Geocoding

Georeferencing and geocoding are different but similar processes since both involve fitting data to the appropriate coordinates of the real world. Georeferencing is the process of assigning coordinates to vectors or rasters so they can be oriented accurately on a model of the Earth's surface. The data used in geocoding are addresses and location descriptors (city, country, etc.). Each of these locations is given the exact coordinates of reference for that location on the surface of the Earth.

Using Spatial Data

The most common way that spatial data is processed and analyzed is using a GIS, or, geographic information system. These are programs or a combination of programs that work together to help users make sense of their spatial data. This includes management, manipulation and customization, analysis, and creating visual displays. A user will typically use multiple spatial datasets at one time and compare them or combine them with one another. Each spatial dataset may be referred to as a layer.

If you were using GIS for a municipality project, you might have vector data like street data (lines), neighbourhood boundary data (polygons), and high school locations (points). Each dataset would exist as its own layer in your GIS. Placement of layers is important for visual purposes as it will help you understand the various types of data and present your findings in an easily understandable way. In this case, you would want to make sure that high school points and street lines are layers above neighbourhood boundaries. Otherwise, you would not be able to see them.

The field and study of GIS extends much further than digital mapping and cartography. It consists of a variety of categories including spatial analysis, remote sensing, and geovisualization. In these GIS fields, the spatial data becomes much more complex and difficult to use.

In addition to raster and vector data, there is also LiDAR data (also known as point clouds) and 3D data. LiDAR data is data that is collected via satellites, drones, or other aerial devices. 3D data is data that extends the typical latitude and longitude 2-D coordinates and incorporates elevation and or depth into the data. While complex, this data is rich with information and can be used to solve a variety of problems pertaining to the Earth's surface.

Using Spatial Data for Graphics

Maps are a common practice of presenting spatial data as they can easily communicate complex topics. They can help validate or provide evidence for decision making, teach others about historical events in an area, or help provide an understanding of natural and human-made phenomena.

When creating visuals, graphics, or maps with spatial data, there are a variety of geographic elements to consider. One of the most important and coincidentally most problematic elements is projection. The projection of a map describes the way that the Earth's surface, a three-dimensional shape, is flattened and presented on a two-dimensional surface. No projection is perfect and depending on your projection you may be sacrificing accuracy in shape, area, distance, or direction.

The City of Vancouver is presented in each of these different projection types. Image a. is using the projection UTM83-10 which is the standard projection used for displaying the City of Vancouver. Image b. is projected using CANBC-Poly resulting in a slightly rotated version of image a. Image c. is projected using LLWGS-84 and is distorted in shape.

a.

b.

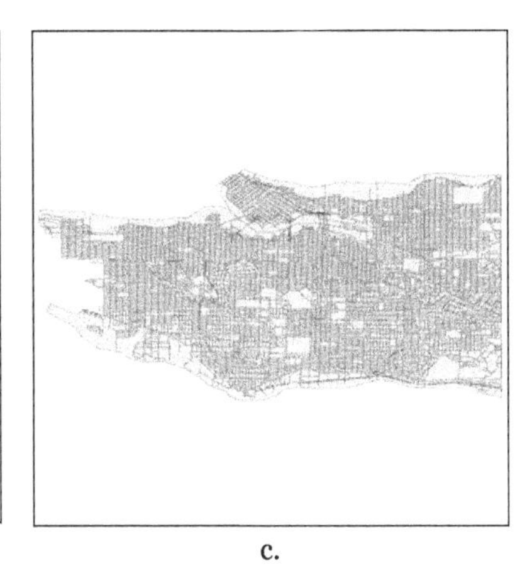

c.

Maps can also be used to present what are typically non-visual elements of society. For example, the occurrence of certain events, income level, any demographic descriptor, or relationships like the number of heat strokes in an area compared to temperature. A simple display method is a classification map, also known as a choropleth map.

Choropleth maps easily communicate differences, consistencies, or patterns across space. Classified areas in a choropleth map will have distinct boundaries whereas heat maps, which demonstrate the concentration or density of a phenomenon, have indistinct boundaries. Classification or heat maps can be used as the bottom layer for other variables like car accidents or crime to highlight certain trends and potential correlations.

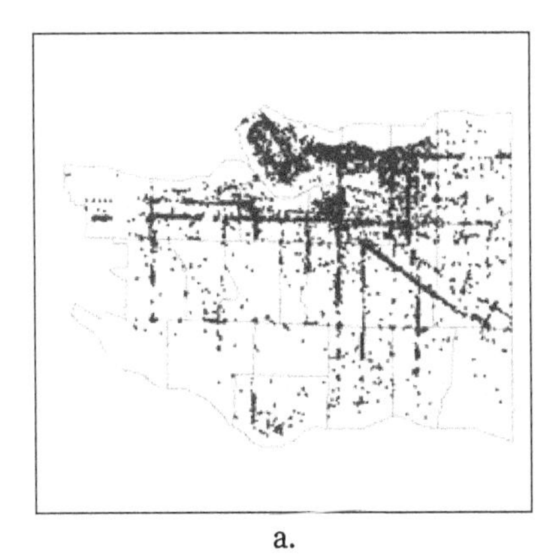

a.

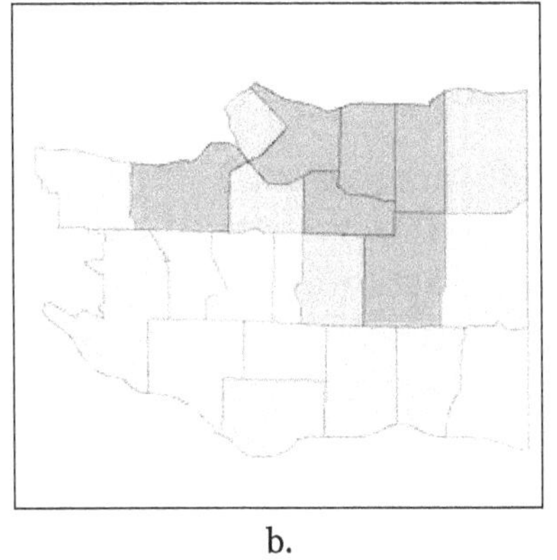

b.

c.

The images above demonstrate a few different ways that spatial data can be displayed. Image a. shows the locations where graffiti has been identified by city custodians in the City of Vancouver. Image b. uses the same point location data, but displays the information as a choropleth map. City region boundaries are highlighted in different colours to describe the density or amount of graffiti taking place within these neighbourhoods. Image c. used the original point location dataset to create a heat map. In this case, city regions are not of interest and rather the spread or pattern of graffiti occurrences throughout the city as a whole is shown.

Using Spatial Data for Statistics

As it is with any data, to truly make sense of spatial data and understand what it is saying you must perform some level of statistical analysis. These processes will help you uncover answers and lead you to make better decisions for your organization. The major difference between spatial data and all other types of data when it comes to statistical analysis is the need to account for factors like elevation, distance, and area in your analytical process.

While needing to account for additional variables about a location may be intimidating, many spatial statistic processes are quite similar to basic statistical methods. For example, interpolation can help you estimate or predict the value of a sample, and spatial interpolation can help you estimate or predict the value of a variable in a sample location. Similarly, spatial autocorrelation measures the degree of similarity between sample locations just like typical autocorrelation is done.

Additional Types of Spatial Data

While spatial data has long been used for analyzing and presenting the Earth's surface, it is not limited to the outdoor environment. There are many architectural, engineering, and construction (AEC) companies that use CAD (computer-aided design) and BIM (building information model) data in their day-to-day activities. While CAD and BIM may not necessarily be thought of as traditional spatial data, they and other AEC formats also need to consider many spatial elements to understand their work.

Mapping is also no longer limited to the natural world. Indoor mapping and wayfinding are becoming much more popular especially in large buildings and institutions like malls, arenas, hospitals, and campuses. This field of study is new but shows no signs of stopping. Everyone has a smartphone these days and uses it to help them navigate the natural world, so why not help people navigate the indoors too?

FME and Spatial Data

FME for Spatial Data Integration

While there are many tools and software that can help you make use of spatial data, FME is the software of choice for those that need to integrate their spatial data. Safe Software and FME came into existence because of this exact problem. Spatial data varies widely and is often stuck in formats that cannot be easily used by all applications, making it extremely difficult for GIS experts to make use of all the information they have. While it was possible to transform proprietary formats in the past, much of the data would be lost in the conversion. Thus, FME was born.

FME is recognized as the data integration platform with the best support for spatial data worldwide. However, it can handle much more than just spatial data and can be easily used by IT and business professionals. FME supports 450+ formats which makes it a flexible data integration tool for those dealing with a large variety of data formats.

Safe Software, the makers of FME, are leaders in the technology world that strive to stay one step ahead of the data integration trends. FME is continuously upgraded to ensure it has been adapted to support new data formats, updated versions of data formats, and large amounts of data. Gone is the idea that individual departments must work in their data silos, with IT structures limiting the company's potential to truly work as one. Data should be able to flow freely no matter where, when, or how it's needed.

SPATIAL DATABASE

Spatial databases provide a strong foundation to accessing, storing and managing your spatial data empire. A database is a collection of related information that permits the entry, storage, input, output and organization of data. A database management system (DBMS) serves as an interface between users and their database.

A spatial database includes location. It has geometry as points, lines and polygons. GIS combines spatial data from many sources with many different people. Databases connect users to the GIS database.

For example, a city might have the waste water division, land records, transportation and fire departments connected and using datasets from common spatial databases. By default, spatial vector features are always associated with non-spatial attribute tables in a GIS. Spatial features store where objects are located on a map. Non-spatial attribute tables explain what the objects on the map represent. Attribute tables are similar to spreadsheets.

Unqid	Page Name	Deer	Roads	Water	Camp
1	A1	9	0	30	0
2	A2	7	0	25	0
3	A3	9	0	15	0

Attribute table example.

FIELDS have different types such as text (*strings*), integers (*whole numbers*) and dates (YYYY/MM/DD). Field name are the title of the column names. A field name should be descriptive of the information being entered in the column. For example, the average population may have a field name of *AVG_POP*.

ROWS in an attribute table represents a spatial feature in the data set or an associated record of that data set.

Rows in an attribute can have relationships with spatial features. There are three types of relationships

- One-to-one relationship.
- One-to-many relationship.
- Many-to-many relationship.

One-to-one Relationships Tie One Table with One Table

What are one-to-one relationships in GIS? Spatial data on a map is always linked to a row in a table. When one feature is linked to one entry in a table, this is called a one-to-one relationship (1-1). For example, here are the geographic locations of five cities in the United States:

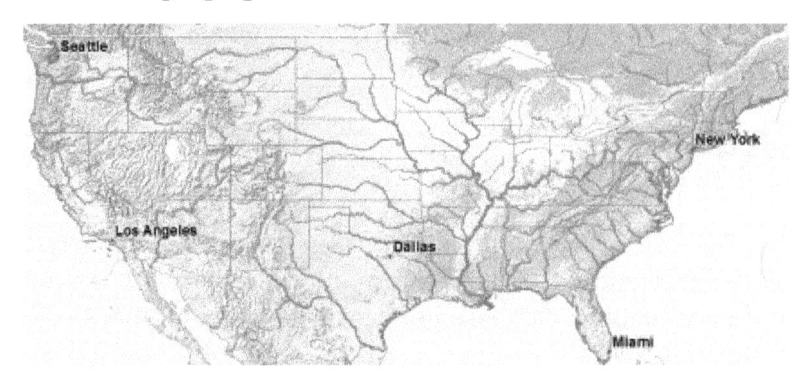

Here is the associated attribute table with these five locations:

Name	Latitude	Longitude	Pop_min	Pop_max
Seattle	47.570002	-122.339985	569369	3074000
New York	40.749979	-73.980017	8008278	19040000
Miami	25.787611	-80.224106	382894	5585000
Los Angeles	33.989978	-118.179981	3694820	12500000
Dallas	32.820024	-96.840017	1211704	4798000

We can get somc extra information about these five cities. We learn their latitude and longitude.

pop max (which is for the metropolitan areas) and pop_min (which is for the incorporated city of the same name).

One-to-many Relationships Ties Multiple Records from a Table Together

When one spatial feature is associated with multiple records in an attribute table this is called a one-to-many relationship (1-M). In this case there is a unique identifier in the spatial table that is used to uniquely identify each row in a table. These are called keys and they exists in both the spatial and non-spatial attribute table.

This unique ID (primary key) is the primary linkage between geographic data and attribute table. This key must exist in the related table as a foreign key. 1-M relationships are commonly set up in spatial databases. Relationship classes set up the type of relationship and which features are being connected by a unique ID. Here is an example where surveys were done at two mountain locations over a period of 3 years. The coordinates of the mountain surveys remained the same over the years:

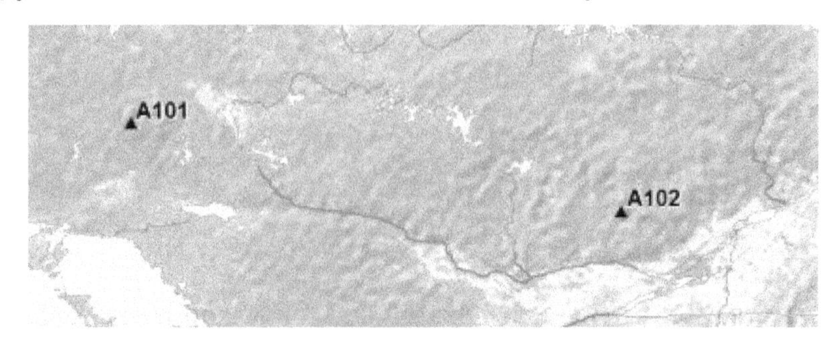

These two mountains have unique IDs of A101 and A102 as shown in their attribute tables:

ID	Lat_Y	Long_x	Name	Featurecla
A101	47.33631	-80.75768	Ishpatina Ridge	Mountain
A102	46.258673	-74.573232	Mont Tremblant	Mountain

Over a period of three years, there have been multiple observations of different animals. The two sites of these mountains have the same IDs – *A101* and *A102*.

The observations at each site can be found within the *OBS* field. The *YEAR* field identifies which year the animal was observed. The two fields that link up these attribute tables are the primary keys *ID* and *SURV_ID*.

SURV_ID	OBS	YEAR
A101	Bighorn sheep	2005
A101	Mountain Goat	2006
A102	Bighorn sheep	2004
A102	Cougar	2005
A102	Elk	2005
A102	Mountain Goat	2006

You can see how we are reducing redundancy. We are no longer storing the latitude and longitude in each observation. This is because each latitude and longitude is only being stored once in the survey points table.

Many-to-many relationships in GIS are much less common in GIS. This type of relationship exists when there are many entries in related attribute tables and vice versa. An intersection table is usually generated to associate many records together.

Relational Database Management System (RDBMS)

The two fields that link up these attribute tables are the primary keys ID and SURV_ID. What ties these tables together is user setting up a 1-M relationship.

This is an example of a RDBMS.

RDBMS is a relational storage concept for data. A relational database contains tables that holds records. Each record holds fields containing data for that record.

It uses normalization to separate tables and link tables together. Normalization doesn't waste space. It breaks out information into discrete components. RDBMS reduces repetitive information. In the example above, we take out duplicate information into multiple tables. There's a foreign key on one table and we can gain access to that table through a relationship.

RDBMS can get very complex such as the example below:

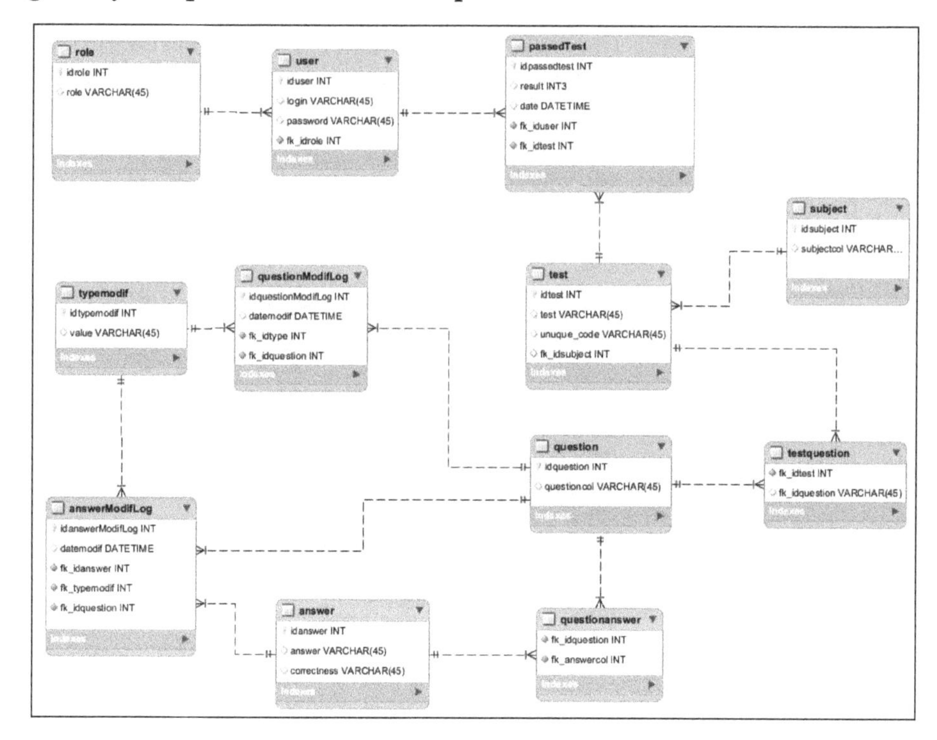

What is Boolean Algebra?

Boolean algebra are conditions used to select features with a set of algebraic conditions. Conditions include AND, OR or NOT.

When you filter the displayed features in the map by setting up a definition query using Boolean algebra. For example: CITY = "LARGE" AND COUNTRY = "NORTH AMERICA" would filter out all the records that meet this criteria

In GIS, Boolean algebra are conditions used to select features with a set of algebraic conditions. Conditions include AND, OR or NOT. Venn diagrams are often used to represent Boolean operations. The name Boolean algebra originated by founder George Boole in 1847.

Structured Query Language (SQL) is how users can interact with the database using Boolean algebra.

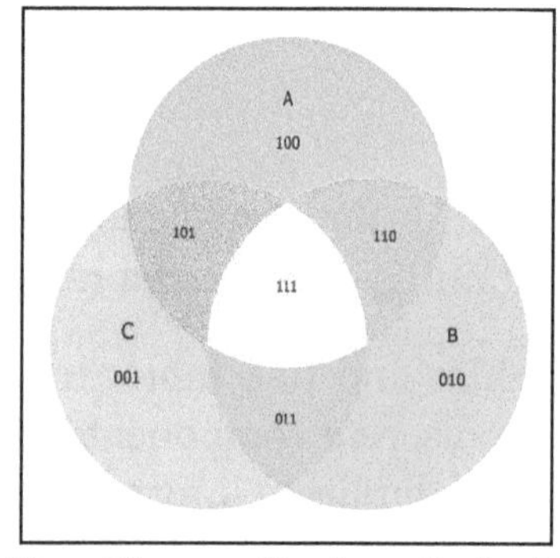

Venn Diagram (Boolean Algebra).

Spatial Database Examples

- Proprietary Esri File Geodatabases stores vectors, rasters, tables, topology and relationships. Schemas can be set up for data integrity. File geodatabases offer structural, performance and data management advantages.
- Open source PostGIS adds spatial objects to the cross-platform PostgreSQL database. The three features that PostGIS delivers to PostgreSQL DBMS are spatial types, indexes and functions. With support for different geometry types, the PostGIS spatial database allows querying and managing information about locations and mapping.
- Other database examples include SQL Server (where geometry is just another data type, like char and int) and Microsoft Access (known as a personal geodatabase in ArcGIS).

Spatial databases provide a mechanism for multiple users to simultaneously access shared spatial data – similar to a DBMS.

The Final Word on Spatial Databases

A database management system (DBMS) allows users to store, insert, delete and update information in a database. RDBMS takes it a step further. It reduces redundancy through normalization. It links tables together through primary and foreign keys.

Spatial databases go a step even further because it records data with geographic coordinates. From geodatabases to PostGIS, spatial databases have quickly become the primary method of managing spatial data.

SPATIAL ANALYSIS

Spatial analysis is in many ways the crux of a GIS, because it includes all of the transformations, manipulations, and methods that can be applied to geographic data to turn them into useful information.

While methods of spatial analysis can be very sophisticated, they can also be very simple. The approach this course will take is to regard spatial analysis as spread out along a continuum of sophistication, ranging from the simplest types that occur very quickly and intuitively when the eye and brain look at a map, to the types that require complex software and advanced mathematical knowledge.

There are many ways of defining spatial analysis, but all in one way or another express the fundamental idea that information on locations is essential. Basically, think of spatial analysis as "a set of methods whose results change when the locations of the objects being analyzed change."

For example, calculating the average income for a group of people is not spatial analysis because the result doesn't depend on the locations of the people. Calculating the center of the United States population, however, is spatial analysis because the result depends directly on the locations of residents.

Types of Spatial Analysis

Types of spatial analysis vary from simple to sophisticated. In this course, spatial analysis will be divided into six categories: queries and reasoning, measurements, transformations, descriptive summaries, optimization, and hypothesis testing.

Queries and reasoning are the most basic of analysis operations, in which the GIS is used to answer simple questions posed by the user. No changes occur in the database and no new data are produced.

Measurements are simple numerical values that describe aspects of geographic data. They include measurement of simple properties of objects, such as length, area, or shape, and of the relationships between pairs of objects, such as distance or direction.

Transformations are simple methods of spatial analysis that change data sets by combining them or comparing them to obtain new data sets and eventually new insights. Transformations use simple geometric, arithmetic, or logical rules, and they include operations that convert raster data to vector data or vice versa. They may also create fields from collections of objects or detect collections of objects in fields.

Descriptive summaries attempt to capture the essence of a data set in one or two numbers. They are the spatial equivalent of the descriptive statistics commonly used in statistical analysis, including the mean and standard deviation.

Optimization techniques are normative in nature, designed to select ideal locations for objects given certain well-defined criteria. They are widely used in market research, in the package delivery industry, and in a host of other applications.

Hypothesis testing focuses on the process of reasoning from the results of a limited sample to make generalizations about an entire population. It allows us, for example, to determine whether a pattern of points could have arisen by chance based on the information from a sample. Hypothesis testing is the basis of inferential statistics and forms the core of statistical analysis, but its use with spatial data can be problematic.

Uncertainty in the Conception of Geographic Phenomena

Many spatial objects are not well defined or their definition is to some extent arbitrary, so that people can reasonably disagree about whether a particular object is x or not. There are at least four types of conceptual uncertainty.

Spatial Uncertainty

Spatial uncertainty occurs when objects do not have a discrete, well defined extent. They may have indistinct boundaries (where exactly does a wetland end?), they may have impacts that extend beyond their boundaries (should an oil spill be defined by the dispersion of pollutants or by the area of environmental damage?), or they may simply be statistical entities. The attributes ascribed to spatial objects may also be subjective—for example, the spatial distributions of poverty and biodiversity depend on human interpretations of what these things mean.

Vagueness

Vagueness occurs when the criteria that define an object as x are not explicit or rigorous. In a land cover analysis, how many oaks (or what proportion of oaks) must be found in a tract of land to qualify it as oak woodland? What incidence of crime (or resident criminals) defines a high crime neighborhood?

Ambiguity

Ambiguity occurs when y is used as a substitute, or indicator, for x because x is not available. The link between direct indicators and the phenomena for which they substitute is straightforward and fairly unambiguous. Soil nutrient levels (y) are a direct indicator of crop yields (x). Indirect indicators tend to be more ambiguous and opaque. Wetlands (y) are an indirect indicator of animal species diversity (x). Of course, indicators are not simply direct or indirect; they occupy a continuum. The more indirect they are, the greater the ambiguity and the less certain it is that an object being approximated using y really is x.

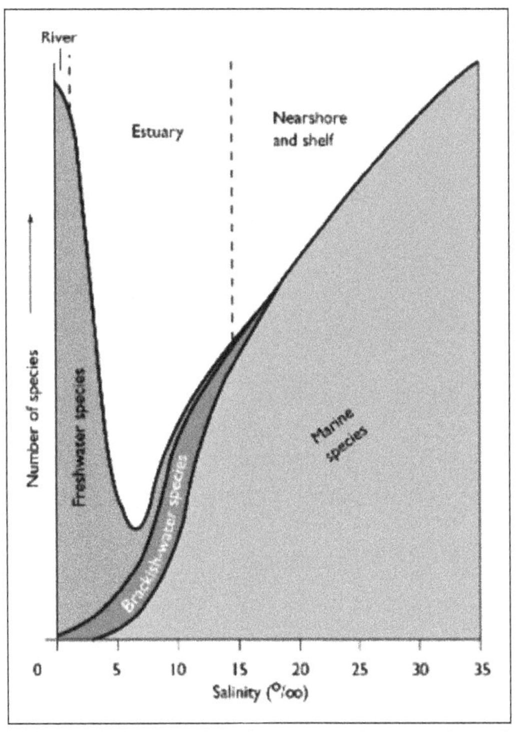

Salinity (x) as a direct, unambiguous indicator of number of species (y), freshwater and marine.But could you correctly estimate the ocean salinity, just from the number of species? Quite ambiguous.

Regionalization Problems

Regional geography is largely founded on the creation of a mosaic of zones that make it easy to portray spatial data distributions. A uniform zone is defined by the extent of a common characteristic, such as climate, landform, or soil type. Functional zones are areas that delimit the extent of influence of a facility or feature—for example, how far people travel to a shopping center or the geographic extent of support for a football team.

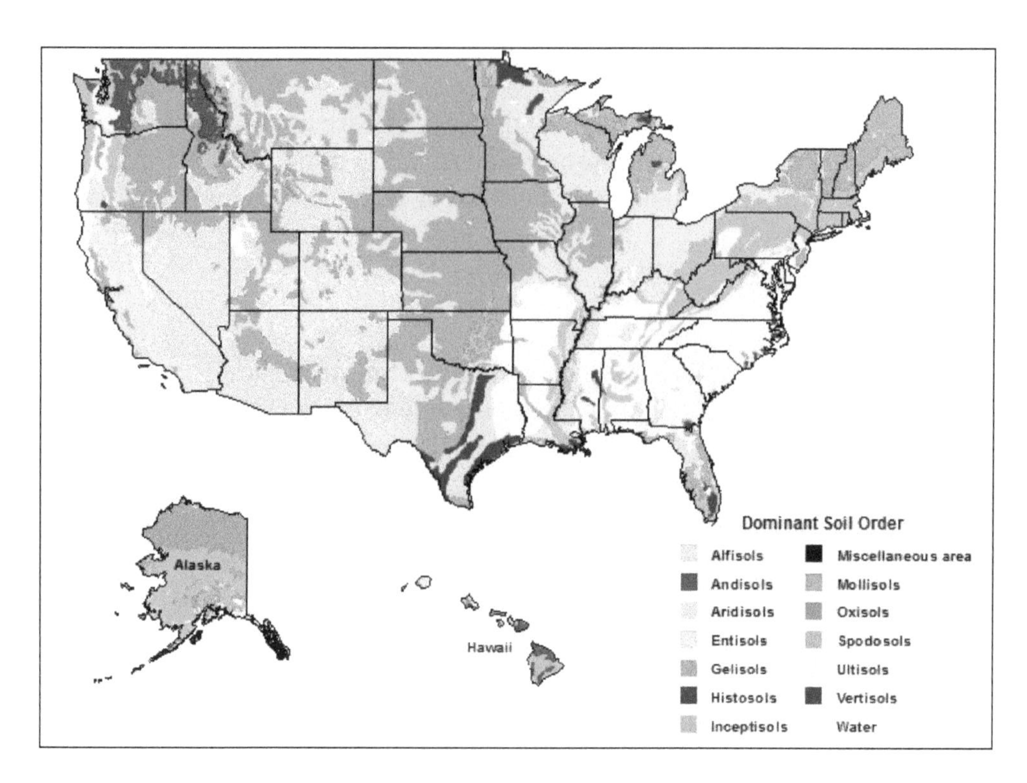

Regionalization problems occur because zones are artificial. In the development of climate zones, for instance, experts may disagree on what combination of characteristics defines a zone, how these characteristics should be weighted to create a composite indicator, and what the minimum size threshold for a zone is. This should not be surprising: after all, spatial distributions tend to change gradually, while zones imply that there are sharp boundaries between them.

Uncertainty in the Measurement of Geographic Phenomena

Error occurs in physical measurement of objects, in the recording of socioeconomic attributes, and in digital data capture. This error creates further uncertainty about the true nature of spatial objects.

Physical Measurement Error

Instruments and procedures used to make physical measurements are not perfectly accurate. For example, a survey of Mount Everest might find its height to be 8,850 meters, with an accuracy of plus or minus 5 meters.

In addition, the earth is not a perfectly stable platform from which to make measurements. Seismic motion, continental drift, and the wobbling of the earth's axis cause physical measurements to be inexact.

Digitizing Error

A great deal of spatial data has been digitized from paper maps. Digitizing, or the electronic tracing of paper maps, is prone to human error. Lines may be drawn too far, not far enough, or missed entirely. Errors caused by digitizing mistakes can be partially, but not completely, fixed by software.

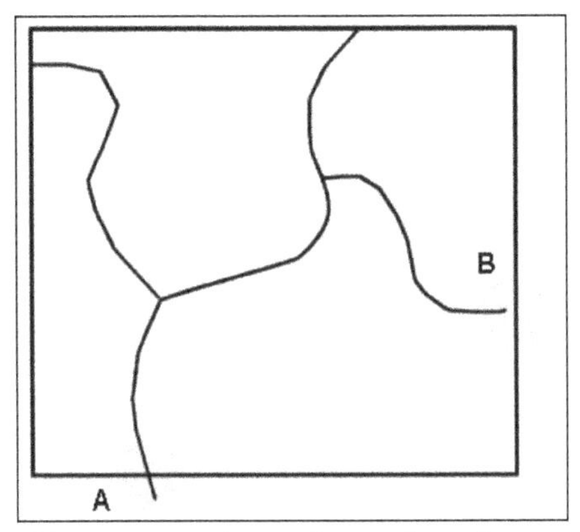

Line segment A overshoots the polygon boundary. Line segment B undershoots it.

Additional error occurs because adjacent data digitized from different maps may not align correctly. This problem can also be partially corrected through a software technique called rubbersheeting.

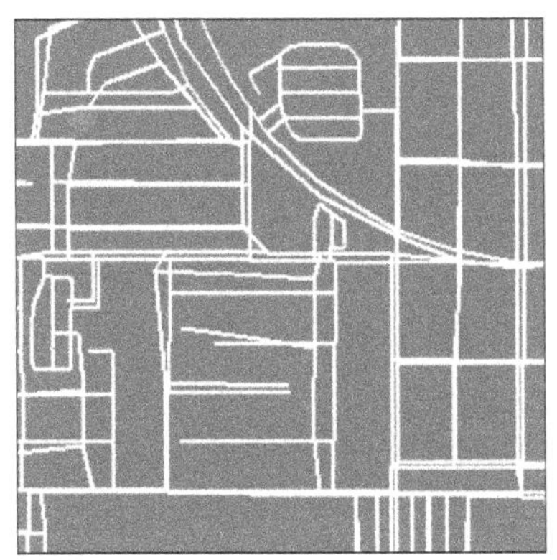

Two data sets representing the same streets do not align with each other. One set can be aligned with the other by a systematic transformation of coordinates called rubbersheeting.

Error Caused by Combining Data Sets with Different Lineages

Data sets produced by different agencies or vendors may not match because different processes were used to capture or automate the data. For example, buildings in one data set may appear on the opposite side of the street in another data set.

Two street data sets for part of Goleta, California, USA. The red and green lines failto match by as much as 100 meters.

Error may also be caused by combining sample and population data or by using sample estimates that are not robust at fine scales. "Lifestyle" data are derived from shopping surveys and provide business and service planners with up-todate socioeconomic data not found in traditional data sources like the census. Yet the methods by which lifestyle data are gathered and aggregated to zones as compared to census data may not be scientifically rigorous.

Uncertainty in the Representation of Geographic Phenomena

Representation is closely related to measurement. Representation is not just an input to analysis, but sometimes also the outcome of it. For this reason, we consider representation separately from measurement.

Uncertainty in the Raster Data Structure

The raster structure partitions space into square cells of equal size (also called pixels). Spatial objects x, y, and z emerge from cell classification, in which Cell A1 is classified as x, Cell A2 as y, Cell A3 as z, and so on, until all cells are evaluated. A spatial object x can be defined as a set of contiguous cells classified as x.

Commonly, a cell is not purely one thing or another, but might contain some x, some y, and maybe a bit of z within its area. These impure cells are termed "mixels." Because a cell can hold only one value, a mixel must be classified as if it were all one thing or another. Therefore, the raster structure may distort the shape of spatial objects.

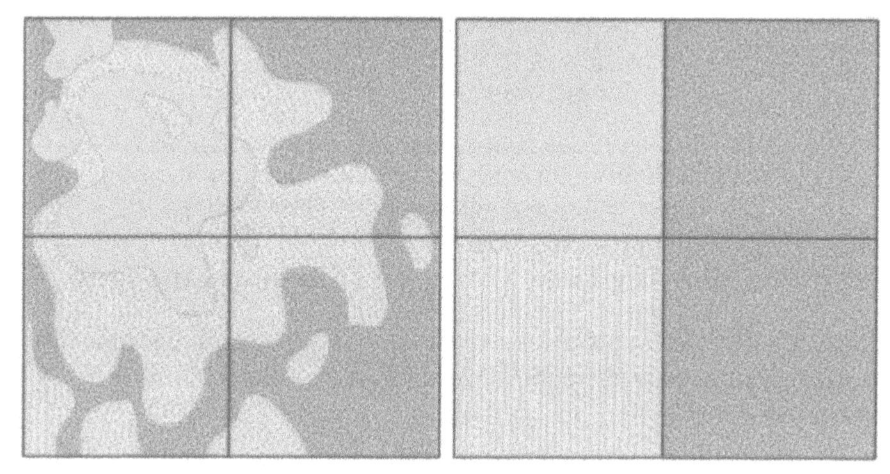

On the left are four mixels; on the right four pixels classified from them. Typically, the pixels will represent the dominant mixel value or the value found at the mixel centroid. Either way, some reality is lost.

Uncertainty in the Vector Data Structure

Socioeconomic data—facts about people, houses, and households—are often best represented as points. For various reasons (to at a zonal level, such as census tracts or ZIP Codes. This distorts the data in two ways: first, it gives them a spatially inappropriate representation (polygons instead of points); second, it forces the data into zones whose boundaries may not respect natural distribution patterns.

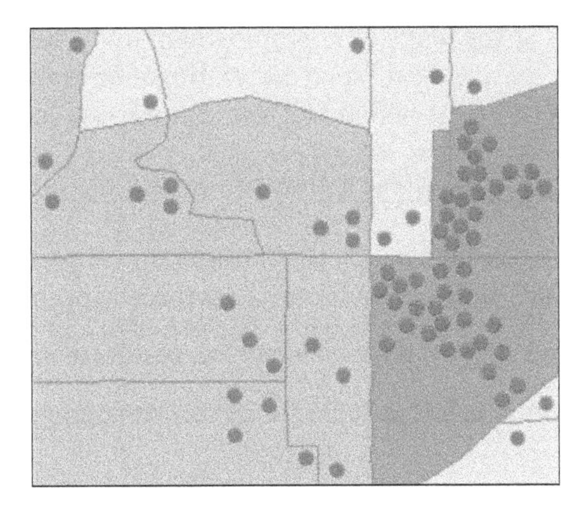

True locations of socioeconomic data (orange points representing households) are often aggregated to zones, such as census tracts, to protect privacy. In this example, two significant distortions occur, neither of which is evident from an examination of the polygon layer by itself. First, points are clustered in corners of polygons, not smoothly distributed as the polygon values imply. Second, some zonal values are based on many data points and others on just a few. The information foundation is not level.

Uncertainty in the Analysis of Geographic Phenomena

Spatial analysis methods can create further uncertainty.

The Ecological Fallacy

The ecological fallacy is the mistake of assuming that an overall characteristic of a zone is also a characteristic of any location or individual within the zone.

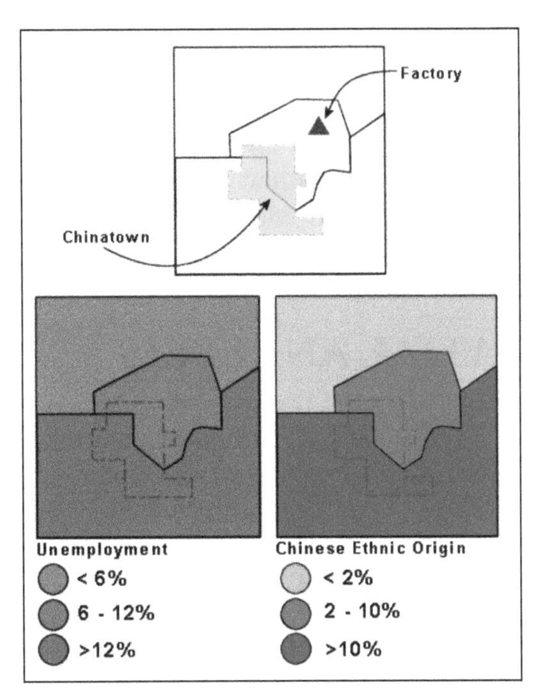

The ecological fallacy. A look at the maps of unemployment and Chinese ethnicity suggests a correlation between them, yet there may be no such correlation. For example, the high unemployment may be caused by the closing of a factory that has few Chinese employees.

Modifiable Areal Unit Problem (MAUP)

The results of data analysis are influenced by the number and sizes of the zones used to organize the data. The Modifiable Area Unit Problem has at least three aspects:

- The number, sizes, and shapes of zones affect the results of analysis.
- The number of ways in which fine-scale zones can be aggregated into larger units is often great.
- There are usually no objective criteria for choosing one zoning scheme over another.

An example of the influence of the number of zones on analysis is the 1950 study by Yule and Kendall which found that the correlation between wheat and potato yields in England changed from low to high as the data were grouped into fewer and fewer zones (starting with 48 and ending with 2).

An example of the influence of zone shape is gerrymandering, in which voting district boundaries are manipulated in order to engineer a desired election outcome.

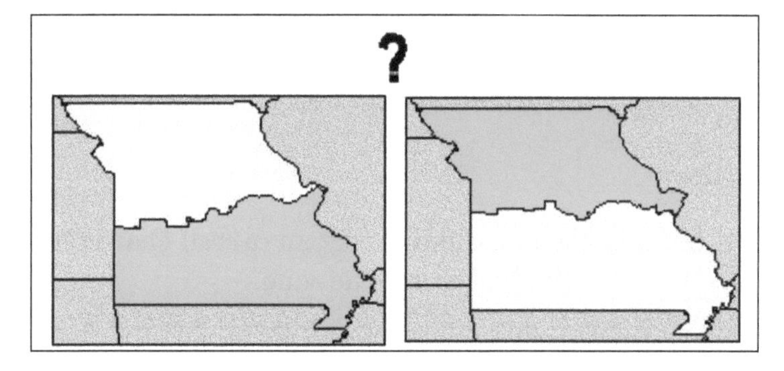

A simple illustration of the MAUP. State of Missouri county data have been aggregated and grouped into one of two zones. A quite minor change in the path of the zonal boundary leads to a different interpretation of whether the northern or southern portion of Missouri has a greater population (darker shade of green represents a higher population).

WORKING OF SPATIAL ANALYSIS

Most data and measurements can be associated with locations and, therefore, can be placed on the map. Using spatial data, you know both what is present and where it is. The real world can be represented as discrete data, stored by its exact geographic location (called "feature data"), or continuous data represented by regular grids (called "raster data"). Of course, the nature of what you're analyzing influences how it is best represented. The natural environment (elevation, temperature, precipitation) is often represented using raster grids, whereas the built environment (roads, buildings) and administrative data (countries, census areas) tends to be represented as vector data.

Further information that describes what is at each location can be attached; this information is often referred to as "attributes."

In GIS each dataset is managed as a layer and can be graphically combined using analytical operators (called overlay analysis). By combining layers using operators and displays, GIS enables you to work with these layers to explore critically important questions and find answers to those questions.

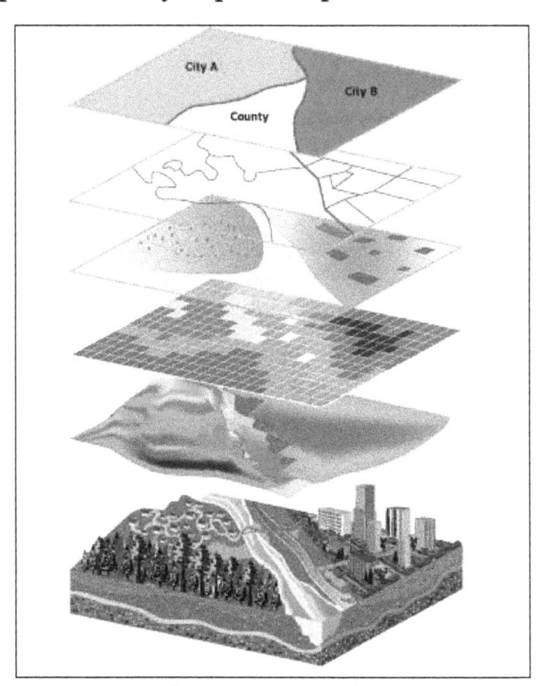

The idea of stacking layers containing different kinds of data and comparing them with each other on the basis of where things are located is the foundational concept of spatial analysis. The layers interlock in the sense that they are all georeferenced to true geographic space.

In addition to locational and attribute information, spatial data inherently contains geometric and topological properties. Geometric properties include position and measurements, such as length, direction, area, and volume. Topological properties represent spatial relationships such as connectivity, inclusion, and adjacency. Using these spatial properties, you can ask even more types of questions of your data to gain deeper insights.

Anatomy of an Overlay Analysis

GIS analysis can be used to answer questions like: Where's the most suitable place for a housing development? A handful of seemingly unrelated factors—land cover, relative slope, distance to existing roads and streams, and soil composition—can each be modeled as layers, and then analyzed together using weighted overlay, a technique often credited to landscape architect Ian McHarg.

The true power of GIS lies in the ability to perform analysis. Spatial analysis is a process in which you model problems geographically, derive results by computer processing, and then explore and examine those results. This type of analysis has proven to be highly effective for evaluating the geographic suitability of certain locations for specific purposes, estimating and predicting outcomes, interpreting and understanding change, detecting important patterns hidden in your information, and much more.

The big idea here is that you can begin applying spatial analysis right away even if you are new to GIS. The ultimate goal is to learn how to solve problems spatially. Several fundamental spatial analysis workflows form the heart of spatial analysis: spatial data exploration, modeling with GIS tools, and spatial problem solving.

Spatial Data Exploration

Spatial data exploration involves interacting with a collection of data and maps related to answering a specific question, which enables you to then visualize and explore geographic information and analytical results that pertain to the question. This allows you to extract knowledge and insights from the data. Spatial data exploration involves working with interactive maps and related tables, charts, graphs, and multimedia. This integrates the geographic perspective with statistical information in the attributes. It's an iterative process of interactive exploration and visualization of maps and data.

Smart mapping is one of the key ways that data exploration is carried out in ArcGIS. It's interesting because it enables you to interact with the data in the context of the map symbology. Smart maps are built around data-driven workflows that generate intelligent data displays and effective default ways to view and interact with your information to see things such as your data's distribution.

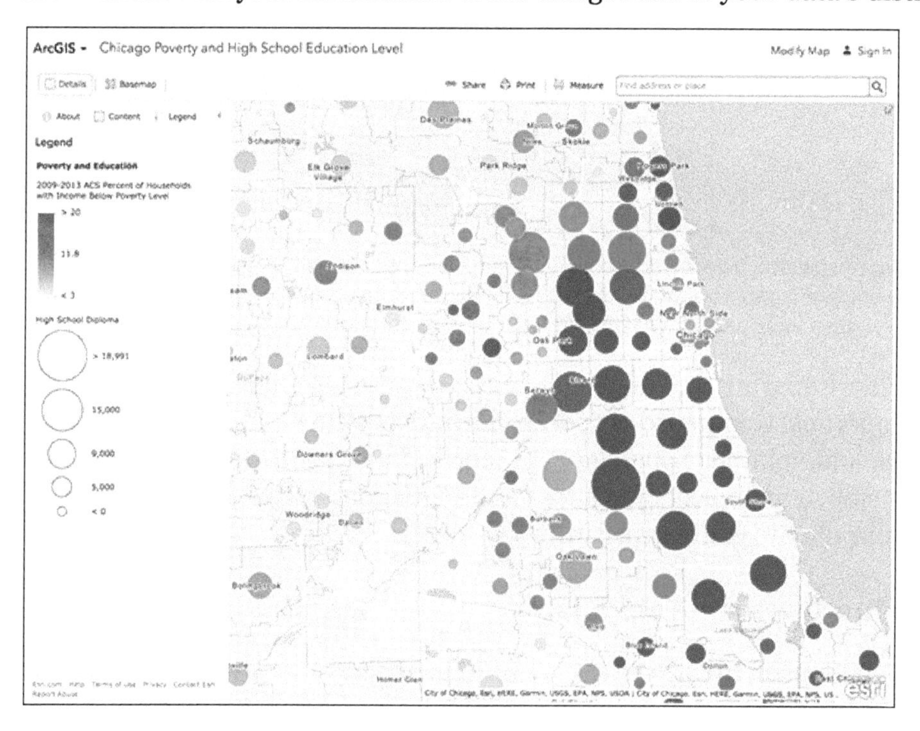

Smart mapping allows you to choose multiple attributes from your data, and visualize the patterns from each attribute within a single map using both color and size to differentiate (also referred to as bivariate mapping). This can be valuable for exploring your data, and allows you to tell a story using one map instead of many.

Combining Interactive Charts and Graphs with GIS Maps

Visualization with charts, graphs, and tables is a way to extend the exploration of your data, offering a fresh way to interpret analysis results and communicate findings. Typically you might begin by browsing through the raw data, looking at records in the table. Then maybe you'd plot (geocode) the points onto the map with different symbology and begin creating different types of charts (bar, line, scatter plot, and so on) to summarize the data in different ways (by district, by type, or by date).

Next, you can begin to examine the temporal trends in the data by plotting time on line charts. Information design is used to arrange different data visualizations to interpret analysis results. Combine a series of your strongest, clearest elements such as maps, charts, and text in a layout that you present and share.

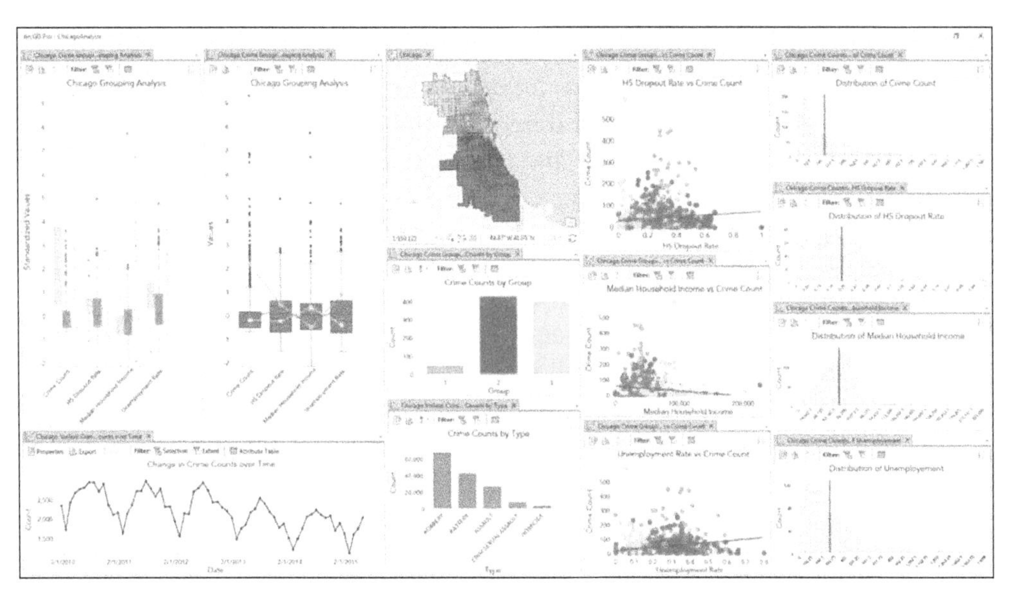

Finding the signal in the noise. Visualizing data through charts helps uncover patterns, trends, relationships, and structure in data that may otherwise be difficult to see as raw numbers. Depicting violent crime statistics from Chicago, a combination of chart and map styles work together to unlock patterns and meaning from what started out as pure tabular data.

SPATIAL DATA REPRESENTATION

GIS technology integrates common spatial database operations such as query and statistical analysis with the unique visualization and geographic analysis benefits offered by maps. Nowadays, evolving with the Internet, the Web-Based GIS, which handles geographic information on the Internet, has appeared and is rapidly evolving as Internet and Web technologies change.

Maps are the primary media of geographical information and the elementary objects manipulated in GIS. They are graphic representations of geographic surfaces on plane, generalized by mathematical rules and represented by visual symbols for different purpose. Maps imply the distributions, states and association of diverse natural or social phenomena. In the existent GIS, almost all of them adopt the layer-based approach to represent geographic information in map. In the layer-based approach, the spatial data are represented in a set of thematic maps, named layer, which denote some given themes such as road, building, subway, contour, border, and so on.

The layer-based approach has an advantage that is easy to process the spatial query and spatial analysis. As we know, the superposition (overlap) of spatial objects is a problem to process in spatial database. The management such as point query processing, region query processing and spatial information analysis processing of overlapped data in spatial objects is difficult to handle because of the too many candidates of spatial object. This problem can be avoid effectively by using the layer-based approach because in general, there is no superposition (overlap) of spatial object in a thematic map of real world.

However, the layer-based approach has the drawback as following:

- It is difficult to extend the map. After the map has been made, if we need to extend the map, well, it is often occurred, such as, for an existent map, when we apply it for a new, special application, we need add corresponding geographic information in the existent map; On the other hand, it is often occurred that when the map is made completely, the new, more detailed geographical information are got, and need to add in the existent map. In these times, it is complex and difficult to add new spatial data in existent map.
- When we display the spatial data in the user screen, there are many spatial objects invisibly because the scale ratio between map and view windows (it seems that we cannot see any building when we look earth in the moon). But for the spatial objects organized in layer must be transferred and displayed. Here the huge size of spatial data transmission affects rapidly the system performance, especially in the Web-Based GIS that processes spatial data in slow Internet environment. We can control it among layers of a map though "controlling display scale mechanism" but cannot prevent it inside one layer.

Layer-based Approach

There are many approaches to represent geographic information in GIS, such as layer-based approach and feature-based approach etc. In the layer-based approach, the spatial data are represented in a set of thematic maps, named layer, which denote some given themes such as road, building, subway, contour, border, and so on. As an example, a map that is composed of layers is shown in figure.

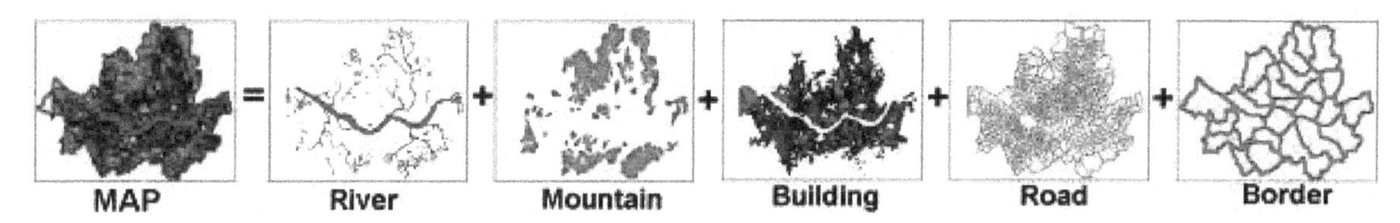

A Sample of Map Composed of Several Thematic Layers.

Generally, the map in layer-based approach is organized as following step:

- Analyzing the special property of target map, determining the theme of layers which will be divided.
- Creating the layers depending on the themes, respectively.
- Creating the indexing data for every layer.

Usually, the organization of map represented in layer-based approach likes as the figure.

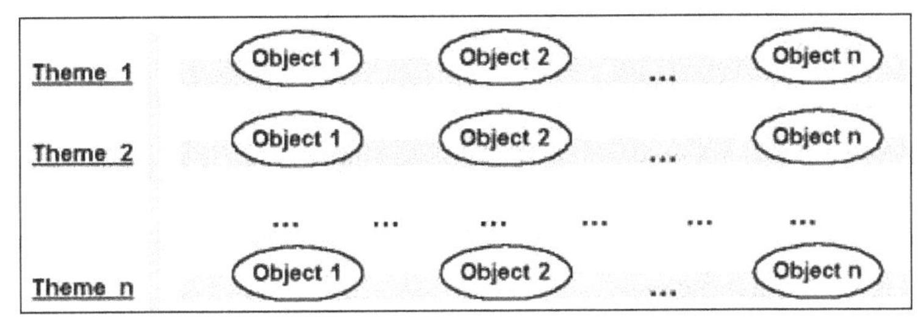

The Organization of Map in Layer-Based Approach.

The layer-based approach has the following advantages:

- Easy query processing and spatial analysis

  The superposition (overlap) of spatial objects is a problem to process in spatial database. The management such as point query processing, region query processing and spatial information analysis processing of overlapped data in spatial objects is difficult to handle because of the too many candidates of spatial object. This problem can be avoid effectively by using the layer-based approach because in general, there is no superposition (overlap) of spatial object in a thematic map of real world.

- Efficient management of data divided-transmission

  When a map is represented in computer devices such as screen, considering display effect in user point of view, it is unnecessary to display all of the details of map. Here the display-scale control is used in the layers of map, and just the necessary layers are displayed. Actually, many traditional desktop GIS and existent Internet GIS hold out the concept of layer in representation of map.

The layer-based approach has the drawback as following:

- It is difficult to extend the map.
- It cannot availably save the question of displaying efficiency about huge size of spatial data in GIS.

The Scale-based Approach

The Principle of Scale-based Approach

In the life, we have the following experience: when we look the object such as a building in the distance of 1 kilometer or just in the front of it, the view of the object in our eyes is different. Although the object exists in the world itself, it is represented in different appearance in our eyes just because of the different distance of our view.

In the representation of spatial data in GIS, we can make use of the principle of vision in people's eyes above. When we browse the Asia map in GIS, we can imagine that we do it in the spacecraft; and when we browse the panorama of Seoul, the capital of Korea, we can imagine that we do it in the plane. In more, when we browse small house in the Seoul, we can imagine that we do it in rooftop of the building that just is in side of it.

In the other hand, the view of map will change belong with the distance of map and our eyes. That is, if we observe a map in a far distance of view, many tiny spatial objects in the map will be invisible whereas the huge size of spatial objects will be invisible when we observe a map in a near distance of view. In this time, it is not necessary for the concerned spatial objects to display. So, for the map in which spatial objects are out of order, we can layer it in several scale-view using the size of spatial objects. That is, we can separate the map the several scale-view depending on the resolutions (scales) of map.

The Three Level Data Model

General, maps are represented in two levels: the set of spatial objects, named layer and spatial object itself. scale-view, block, and spatial object.

Firstly, "Spatial Object' is a basic element about the graphic representation of geographic phenomena in GISs. Here every geographic object is represented in a geometric object, which possesses location, shape, topologic relations and geometric sizing information as figure:

Spatial Object Structure		
	OID	ID of Spatial Object
Spatial Part	MBR	MBR of Spatial Object
Spatial Part	Type	Type of Geometric Object: (Polygon, Polyline, ...)
Spatial Part	Data	Data of Geometric Object
Aspatial Part	Attribute Data	Spatial Object's attribute data
	PBID	ID of parent block

Data Structure of Spatial Object.

"Block" is abstract representation of a set of spatial objects. It can be look as an abstract spatial object (polygon). Inside it, it includes a set of spatial objects named child object, or a set of blocks named child block. So it is also an entity of these spatial objects. The "Block" can be separated into 3 kinds: Pure Block, Mixed Block and End Block. The elements of "Pure Block" are all blocks. The "Mixed Block" not only includes blocks but spatial objects while the "End Block" just includes spatial objects. The type of "Block" is illustrated in figure, and the data structure of Block is illustrated in figure.

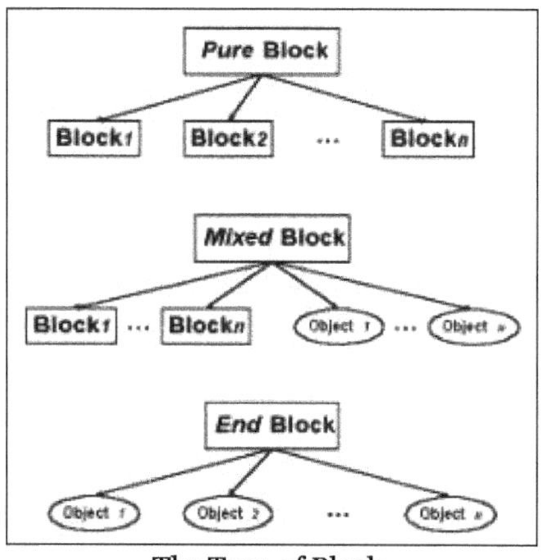

The Type of Block.

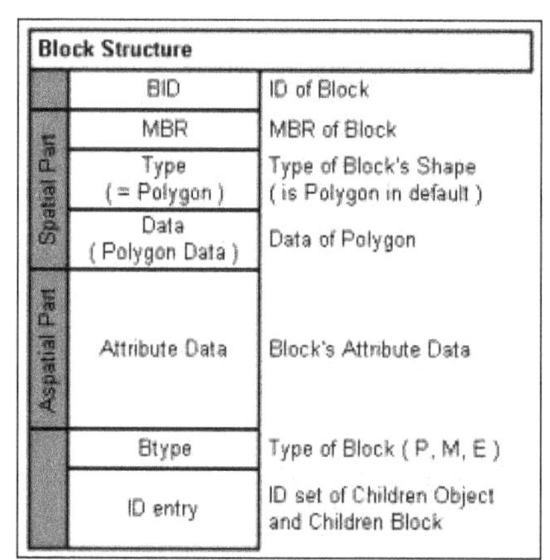

Block Structure		
Spatial Part	BID	ID of Block
Spatial Part	MBR	MBR of Block
Spatial Part	Type (= Polygon)	Type of Block's Shape (is Polygon in default)
Spatial Part	Data (Polygon Data)	Data of Polygon
Aspatial Part	Attribute Data	Block's Attribute Data
	Btype	Type of Block (P, M, E)
	ID entry	ID set of Children Object and Children Block

The Data Structure of Block.

At finally, "Scale-View" is a set of block and a set of spatial objects that associate with a certain given scale. Liking layer concept in layer-based approach, here a complete map is represented in a set of "scale-view", which denote a view of this map in certain resolution (scale).

The Map Structure in Scale-based Approach

In scale-based approach, map is composed of a series of scale-views that include a series of blocks and spatial objects associating a certain given scales. In the scale-view, the block is looked as a common spatial object (polygon) in where it has spatial data and attribute data.

In order to locate the spatial object in given OID (spatial Object ID) or BID (Block ID), it is necessary to add the "spatial object allocation table" to record the relation of ID and storage position of spatial object. In addition, for save information about child object or child block relation of block, the "block link table" is also necessary.

Well, the details of implementation on map structure is not limit to one form. There are many methods to implement the map structure depending on concrete storage strategy and environment. Here one implementation on map structure is given in our prototyped GIS that are built to evaluate the scale-based approach. The structure of map in scale-based approach is illustrated in figure as following:

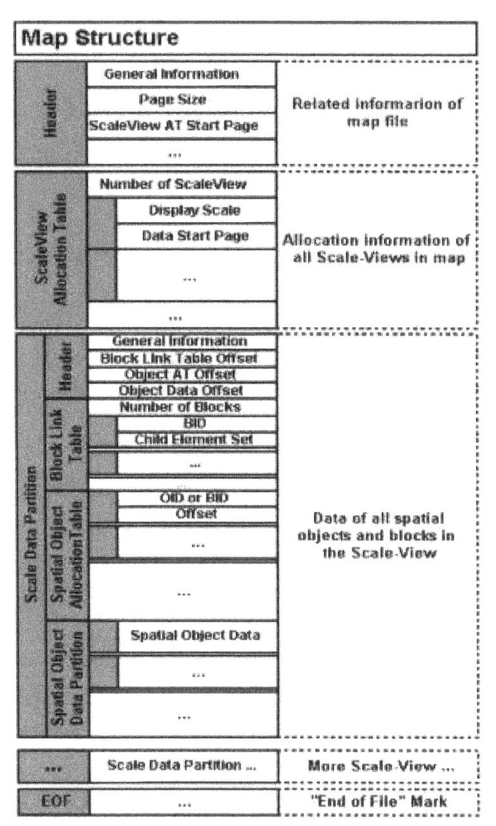

Figure: The File Structure of Map in Scale-Based Approach

Organization of Map in Scale-based Approach

For a map, we organized it in a tree structure depending on a series of resolutions (scales). In the root of tree, it is first scale-view, which just is a "Pure Block", called "root block", the entity of the whole map that represents abstractly the shape of this map as a polygon. And in the second scale-view, we separate the "root block" into several blocks (or a set spatial objects) depending on certain separation rule such as "to separate it in administration boundary". The spatial object that cannot be separated in this rule (such as mountain etc.) just link to parent block as a child object directly. The third scale-view, fourth scale-view are created in so on until the lowest scale is satisfied our request. In this time, all element of the lowest scale-view are spatial objects.

Actually, for extending the map up, we do not create the "root block". Instead of it, we represent the highest scale-view in a set of blocks and a set of spatial object (if there need do it so).

The organization of map in scale-based approach is illustrated in figure:

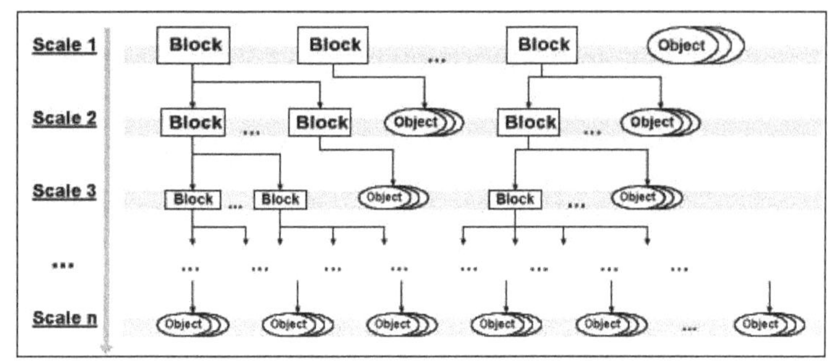

The Organization of Map in Scale-based Approach.

In the organization above, we can find that it is easy to extend map that is represented in scale-based approach. We can extend map in extending the corresponding object in a block and add the new spatial objects in the new created block and link it to the existent map.

Indexing Structure and Extension of Map in Scale-based Approach

As it is well know, a number of spatial indexing structures based MBR of the spatial have been developed such as R-Tree, Cell-Tree, Grid-File and so on. In existent GIS, the most promising one of them is R-Tree and its variations, e.g., R+-Tree and R-Tree. Here we extend the R-Tree indexing structure to use for indexing the map in scale-based approach.

Firstly, for a detail of block, liking as a general map, we index it in R-Tree structure in where the child blocks are treated as general spatial objects. In addition we organize the map indexing structure as a "block link tree" that includes all R-Tree indexing structure of blocks depending on the relation among blocks (child or parent) and the affiliation of block and scale-view. The indexing structure of block and map are illustrated in figure, respectively.

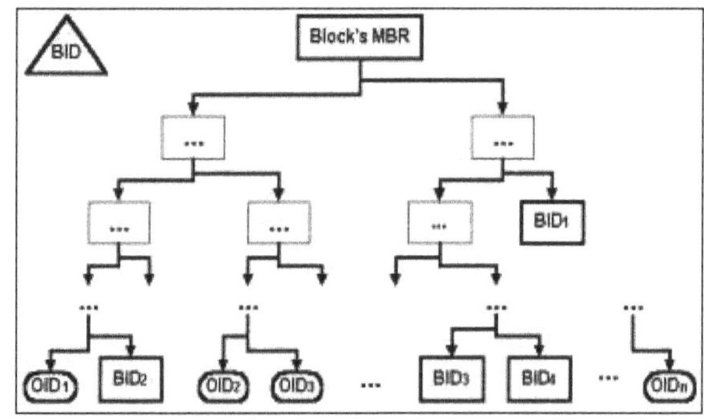

The Indexing Structure of Block.

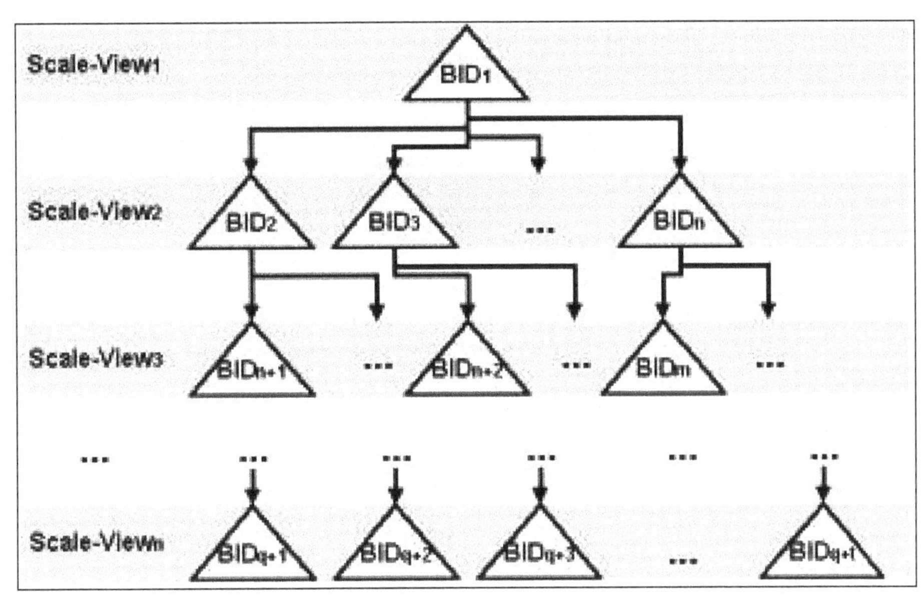

The Indexing Structure of Map.

About the extension of map, there is no problem to join the new spatial objects or blocks to the target map in file storage structure, just only tiny modify on creation of indexing structure.

There are two cases for extension of an existent map, one is extending map more commodiously in order to enlarge scope of map, and another is extending map in more detail in order to increase precision of map.

In first case, for example, there are a map A, has 4 scale-views, A1, A2, A3, A4, and another map B, has 4 coordinate scale-views, B1, B2, B3, B4, if we want join the two map in one map C, the next steps need to be executed:

- Creating index data of map C's root block depending on A's indexing data (A's R*-Tree) and B's indexing structure (B's R*-Tree) (the BMBR of C is new MBR that contains A's MBR and B's MBR);
- Creating Map C's map indexing structure, link the blocks that locate in A1, A2, A3, A4 and B1, B2, B3, B4 (modify the BID of blocks), building coordinate scale-views C1, C2, C3, C4;
- Creating New Map File C, insert map A's map data and B's map data orderly, and modify BID of blocks and OID of spatial objects;
- For every block indexing data (R*-Tree) that are in C1, C2, C3, C4, modifying the related OID to new value that defined in step 3.

In second case, the new extended map can be created just through doing the step 2~4 above because that the scope of extended map (MBR of map) is not changed.

The Prototype System Constitution and Performance Evaluation

In this work, we have created a prototype Web-Based GIS using the proposed spatial data representation. We organize the Korea map in 4 "Scale-View" as exam material. For a sample, a part of the Korea map is listed curtly in figure as following:

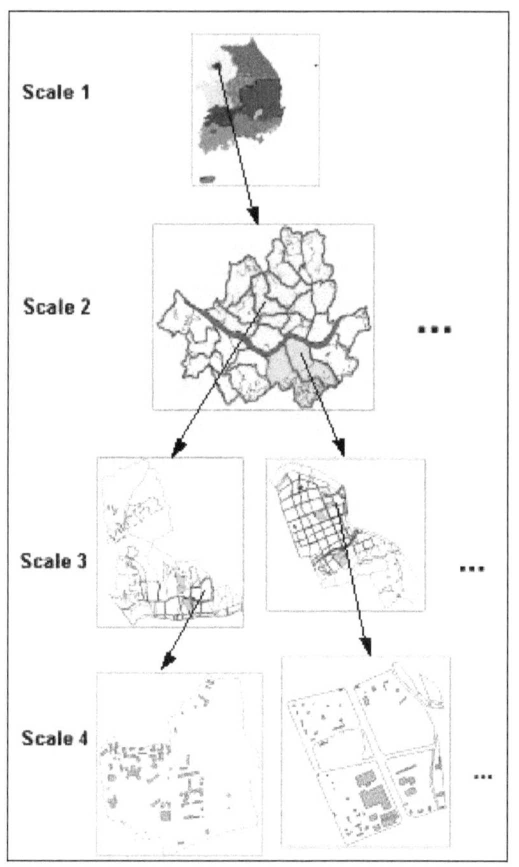

The Sample Map Created in Scale-Based Approach.

In prototype Web-Based GIS, when the user request the all of Korea map, the server just transfers and displays the scale-view of "scale 1", and when the user browses the Seoul map, the server just transmits and displays the scale-view of "scale 2" (or scale-view of "scale 2" and "scale 3" in detail request). Furthermore, when user request street scene of certain district, the server just transmits and displays the corresponding region of scale-view of "scale 4" and so on. Usually, when user request a certain region, the server just transmits and displays the related region of scale-view in "scale 1", "scale 2", "scale 3" or "scale 4" (depending on the size of region) using the map indexing structure.

TOPOLOGY

Topography is the study of the land surface. In particular, it lays the underlying foundation of a landscape. For example, topography refers to mountains, valleys, rivers or craters on the surface.

The origin of topography comes from "topo" for "place" and "graphia" for "writing". It's closely related to geodesy and surveying which are concerned with accurately measuring the land surface. And it's also closely tied to geography and mapping systems like GIS.

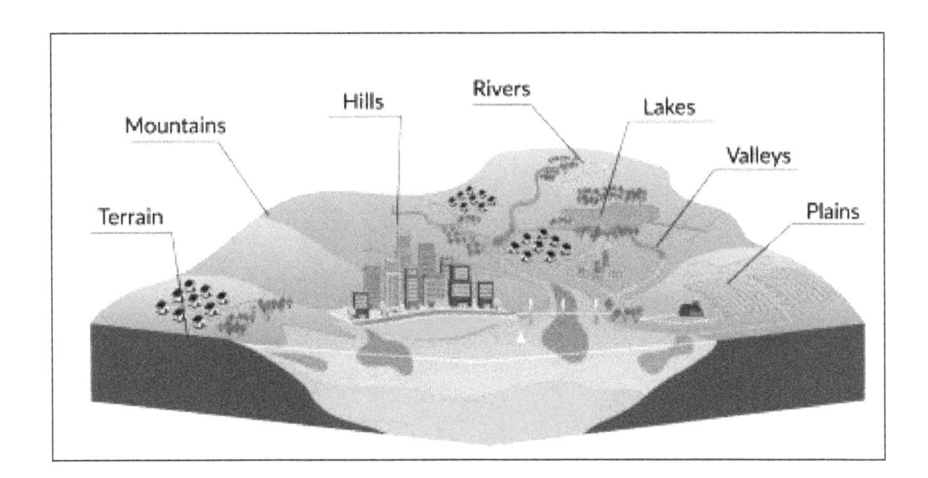

Elevation is the distinguishing factor for topographic maps. In GIS, we use digital elevation models for terrain. Nine out of ten topographic maps show contour lines, which are just lines of equal elevation. The narrow definition of topography is specific to the arrangement of landforms.

But in a broader sense, it incorporates natural and artificial features. For example, topographic maps often tie in administrative boundaries, cities, hydrography, parks, landmarks, transportation and buildings.

Relief and Contours

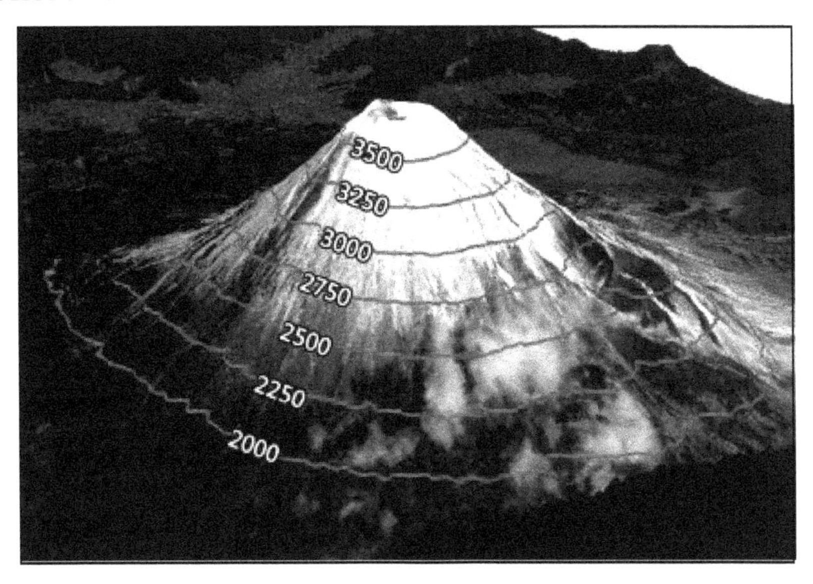

Contour lines (isolines) connect points of equal elevation. By reading contours, we interpret height, slope and shape in topographic maps.

If contours are close together, the slope is steep. But when contours are spread apart, the slope is more gradual.

We use contours for mountains, valleys and bathymetry. For example, Mount Fuji stands at 3,776 meters above sea level. At 250-meter spacing, each contour line represents equal elevations. Almost at the peak of Mount Fuji, it's a 3,750 meter contour line.

Examples of Topography in Maps

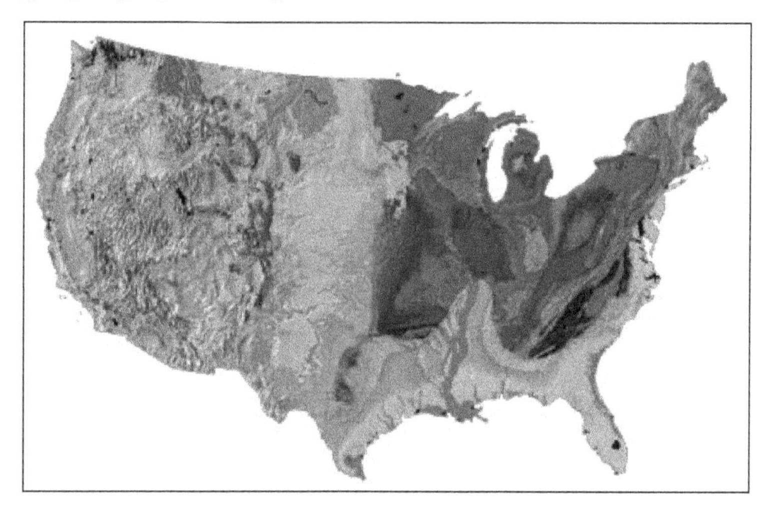

There's no "world authority" in topographic mapping. Instead each country sets their own standards and priorities. Most often the case, each mapping agency designs their topographic maps with a specific goal in mind.

For example, constructing a new highway might drive a topographic map to feature woodland cover, soil types or rock classification along the route. Over time, topographic map series often get periodic updates. But the truth is, they can be complex and take years to create.

In the United States, the first USGS topographic map was surveyed in 1892. Since then, map revisions have continued for over 125+ years. The USGS produce topo maps at 1:250,000, 1:100,000, 1:63,360 and 1:24,000. The most common is the 7.5 minute quadrangle series where one inch in the map represents 24,000 inches on the ground.

Another example of a topographic map is the USGS Tapestry of Time and Terrain. In this colorful map, it overlays topography (hillshade) with underlying rock formations. This helps unravel the geologic history of the continent, such as mountain building events.

Topography Applications and Uses

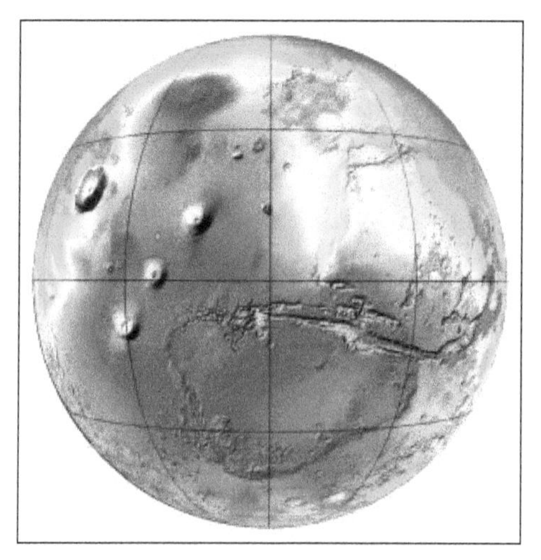

Topographic maps show how rivers flow, how high mountains rise and how steep valleys descend. They lay out the land such as in these examples:

- Engineers use topographic maps to plan a road, construct a cell tower or plan a hydroelectric dam.
- Geologists use topography to understand tectonic activity, landforms and where to dig a mine.
- Hikers use topographic maps to find trails and steepness of slope to plan their ascent.
- Astronomers study the topography outside Earth like on the moon, Mars or an asteroid.
- Climate scientists tie topography into climate models to recognize air and water flow.

As landscapes evolve and technology advances, topographers face an uphill battle for accuracy and completeness.

Topography Analysis

If you want to perform any type of topographic analysis, we highly recommend SagaGIS. It's completely free and open source. Specifically, the Topography Analysis Toolkit is a perfect fit for most types of landscape analysis.

It's so good that you can't find most of these tools in commercial software. In particular, it includes a range of morphometry, hypsometry and other specialty tools. Tools like ruggedness, slope, aspect and curvature can really characterize the terrain. If you want to classify the types of landforms, SagaGIS has out-of-the box tools that do just that. Lastly, we use the topographic position and wetness index to characterize drainage patterns.

Topography is all about location, location, location. Specifically, it's how topography relates to relief. Topography is the arrangement of natural and artificial features in the world. Engineers, geologists and even astronomers use these types of maps for reconnaissance, planning and characterizing the land.

SPATIAL ETL

Spatial ETL, also known as Geospatial Transformation and Load (GTL), provides the data processing functionality of traditional Extract, Transform, Load (ETL) software, but with a primary focus on the ability to manage spatial data (which may also be called GIS, geographic, or map data).

A spatial ETL system may translate data directly from one format to another, or via an intermediate format; the latter being more common when transformation of the data is to be carried out.

Spatial Analysis

Transform

The transformation phase of a spatial ETL process allows a variety of functions; some of these are similar to standard ETL, but some are unique to spatial data.

Spatial data commonly consists of a geographic element and related attribute data; therefore spatial ETL transformations are often described as being either *geometric transformations* - transformation of the geographic element - or *attribute transformations* - transformations of the related attribute data.

Common Geospatial Transformations

- Reprojection: The ability to convert spatial data between one coordinate system and another.
- Spatial transformations: The ability to model spatial interactions and calculate spatial predicates.
- Topological transformations: The ability to create topological relationships between disparate datasets.
- Resymbolisation: The ability to change the cartographic characteristics of a feature, such as colour or line-style.
- Geocoding: the ability to convert attributes of tabular data into spatial data.

Additional Features

Desirable features of a spatial ETL application are:

- Data comparison: Ability to carry out change detection and perform incremental updates.
- Conflict management: Ability to manage conflicts between multiple users of the same data.
- Data dissemination: Ability to publish data via the internet or deliver by email regardless of source format.

- Semantic processing: Ability to understand the rules of different data formats to minimize user input whilst preserving meaning.

Spatial ETL Uses

Spatial ETL has a number of distinct uses:

- Data cleansing: The removal of errors within a dataset
- Data merging: The bringing together of multiple datasets into a common framework - conflation is a good example of this
- Data verification: The comparison of multiple datasets for verification and quality assurance purposes
- Data conversion: Conversion between different data formats.

Although ETL tools for processing non-spatial data have existed for some time, ETL tools that can manage the unique characteristics of spatial data only emerged in the early 1990s.

Spatial ETL tools emerged in the GIS industry to enable interoperability (or the exchange of information) between the industry's diverse array of mapping applications and associated proprietary formats. However, spatial ETL tools are also becoming increasingly important in the realm of Management Information Systems as a tool to help organizations integrate spatial data with their existing non-spatial databases, and also to leverage their spatial data assets to develop more competitive business strategies.

Traditionally, GIS applications have had the ability to read or import a limited number of spatial data formats, but with few specialist ETL transformation tools; the concept being to import data then carry out step-by-step transformation or analysis within the GIS application itself. Conversely, spatial ETL does not require the user to import or view the data, and generally carries out its tasks in a single predefined process.

With the push to achieve greater interoperability within the GIS industry, many existing GIS applications are now incorporating spatial ETL tools within their products; the ArcGIS Data Interoperability Extension being a good example

SPATIAL ANALYSIS IN ARCGIS PRO

Spatial analysis allows you to solve complex location-oriented problems and better understand where and what is occurring in your world. It goes beyond mere mapping to let you study the characteristics of places and the relationships between them. Spatial analysis lends new perspectives to your decision making.

Look at a map of crime in your city and tried to figure out what areas have high crime rates? Explore other types of information, such as school locations, parks, and demographics to try to determine the best location to buy a new home? Whenever we look at a map, we inherently start turning that map into information by analyzing its contents—finding patterns, assessing trends, or making decisions. This process is called spatial analysis, and it's what our eyes and minds do naturally whenever we look at a map.

Spatial analysis is one of the most intriguing and remarkable aspect of GIS. Using spatial analysis, you can combine information from many sources and derive new sets of information by applying a sophisticated set of spatial operators. This comprehensive collection of spatial analysis tools extends your ability to answer complex spatial questions. Statistical analysis can determine if the patterns that you see are significant. You can analyze various layers to calculate the suitability of a place for a particular activity. And by employing image analysis, you can detect change over time. These tools enable you to address critically important questions and decisions that are beyond the scope of simple visual analysis.

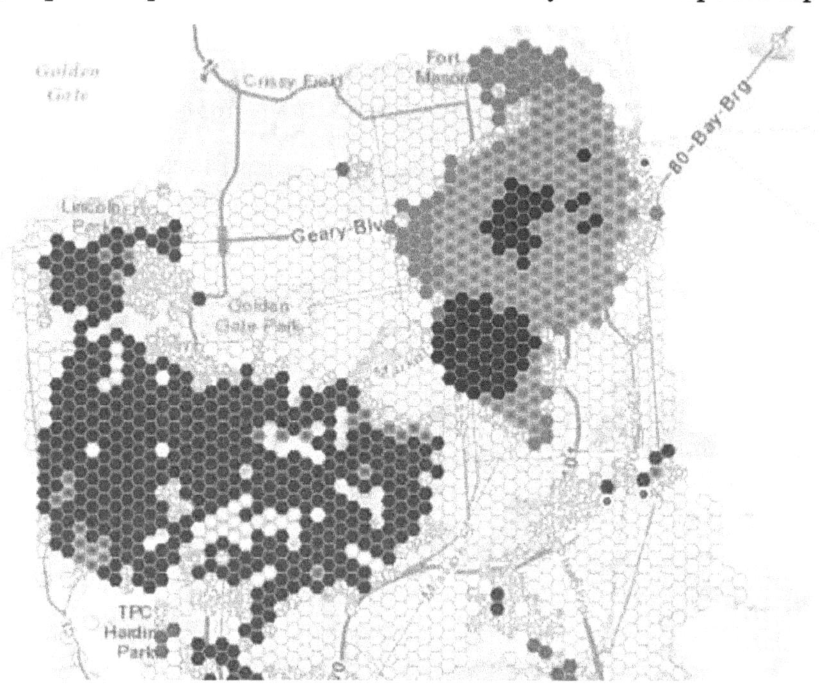

Hot Spot Analysis of San Francisco crimes.

With spatial analysis, you can do the following:

- Determine relationships.
- Understand and describe locations and events.
- Detect and quantify patterns.
- Make predictions.
- Find best locations and paths.

You can use the comprehensive analysis and geoprocessing capabilities in ArcGIS Pro to answer many important spatial questions and perform virtually any type of spatial analysis. To perform spatial analysis, first frame your question, then prepare and analyze your data, and finally visualize and communicate your results. Spatial analysis in ArcGIS Pro is extended from 2D to 3D and also through time.

Work with Geoprocessing Tools

Using geoprocessing tools in ArcGIS Pro, you can perform the following types of operations on geographic data:

- Extract and overlay data.
- Add and calculate attribute fields.
- Summarize and aggregate data.
- Calculate statistics.
- Model relationships and discover patterns.

Automate your Work

If ArcGIS Pro does not include a tool that can answer your specific question, you can build your own custom tool. Using geoprocessing tools and other functions as building blocks, you can create simple or complex workflows as models or scripts. Models and scripts help you automate multi-step processes as well as document processing steps so the workflow can be better understood, examined, and refined.

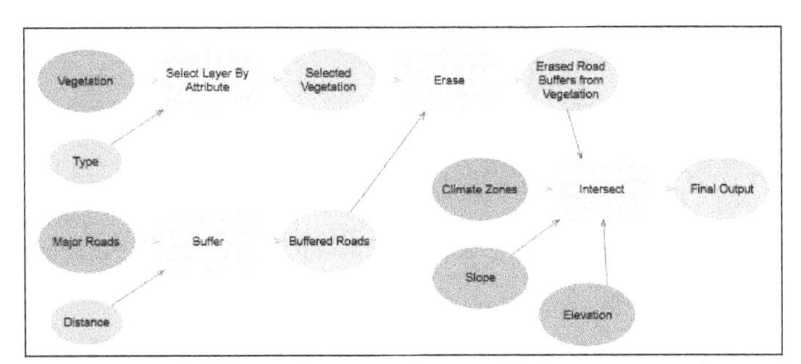

A ModelBuilder model that strings together multiple geoprocessing tools using the output of one tool as the input to another tool is shown.

Visualize your Data in Charts

Visualizing data through charts helps to uncover patterns, trends, relationships, and structure in data that may otherwise be difficult to see as raw numbers in a table. Use charts together with maps to explore your data and help tell a story.

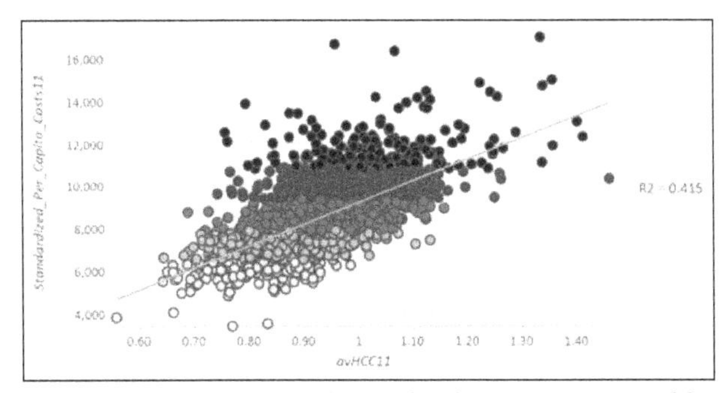

A scatter plot shows the relationship between two variables.

Use Analysis Extensions

ArcGIS Pro provides the following analysis extensions to help you answer even more specialized spatial questions:

- 3D Analyst—Analyze and create 3D GIS data and perform 3D surface operations against rasters, TINs, terrains, and LAS datasets (lidar).
- Business Analyst—Analyze market trends, including customer and competitor analysis, site evaluation, and territory planning.
- Geostatistical Analyst—Analyze and predict the values associated with spatial or spatio-temporal phenomena.
- Image Analyst—Interpret and exploit imagery, perform feature extraction and measurement, and classification and object detection using machine learning.
- Network Analyst—Measure distances and travel times along a network to find a route between multiple locations, create drive-time buffers or service areas, and find the best locations for facilities to serve a set of locations.

Spatial Analyst—Perform interpolation, overlay, distance measurement, density, hydrology modeling, site suitability, and math and statistics on cell-based raster data.

References

- Spatial-data, what-is: safe.com, Retrieved 29 June, 2019
- Spatial-databases: gisgeography.com, Retrieved 30 January, 2019
- How-to-perform-spatial-analysis, analytics, product, products, arcgis-blog: esri.com, Retrieved 31 March, 2019
- What-is-topography: gisgeography.com, Retrieved 14 July, 2019
- Spatial-analysis-in-arcgis-pro, introduction, analysis, help, pro-app: arcgis.com, Retrieved 17 May, 2019

5

Understanding Geocoding

The process of transforming a description to a location on the surface of the Earth is known as geocoding. Some of the major concepts studied in relation to geocoding are geocodes, geocoding addresses, geolocation and reverse geocoding. This chapter has been carefully written to provide an easy understanding of these aspects of geocoding.

GEOCODES

A geocode represents the GPS coordinates of a certain physical location.

Every given location on Earth can be represented by the value of its latitude and longitude, so as long as you can procure these two pieces of information, you can make a geocode to mark that particular spot.

How does One Engage in the Act of Geocoding?

As high-tech and advanced as the handling of geocodes sounds, the process of procuring them can be painstakingly difficult- and in the physical sense, too!

Namely, the first way (and possibly the most reliable one) of acquiring geocodes is to travel to the destination you want to 'capture in code', whip out your GPS device, press some buttons et voila!- you've got yourself a properly captured geocode at the exact location of its 'natural habitat'.

The second way includes less walking and more math. (Which for some people doesn't necessarily represent that much of an improvement.) The process involves mixing up of the two extracted values of a coordinate and then approximating its position by using a certain formula to do the equation.

As the word itself suggests, the second approach revolves around approximation, while the first, spot-on approach leaves very little to no room for error.

Why Geocodes Matter?

One of the biggest problems when it comes to delivering certain goods to a location is that the address is simply not precise enough.

Whether it's through human error, or some other set of unforeseen circumstances, (Such as delivering goods to an enormous facility with multiple entrances.) not being able to find your way toward your destination straight away can cost time and money.

Geocodes have a pinpoint accuracy and can be easily updated when necessary, which makes them a perfect currency for finding any given location.

In conclusion, working with faulty information when delivering goods can result in some unneeded embarrassment, rescheduling and possibly even losing clients! When it comes to pinpointing a place on a map, geocoding is currently the best way to determine any given location and mark it for later reference.

GEOCODE ADDRESSES

Geocodes are the table of addresses. This process requires a table that stores the addresses you want to geocode and an address locator or a composite address locator. This tool matches the addresses against the address locator and saves the result for each input record in a new point feature class.

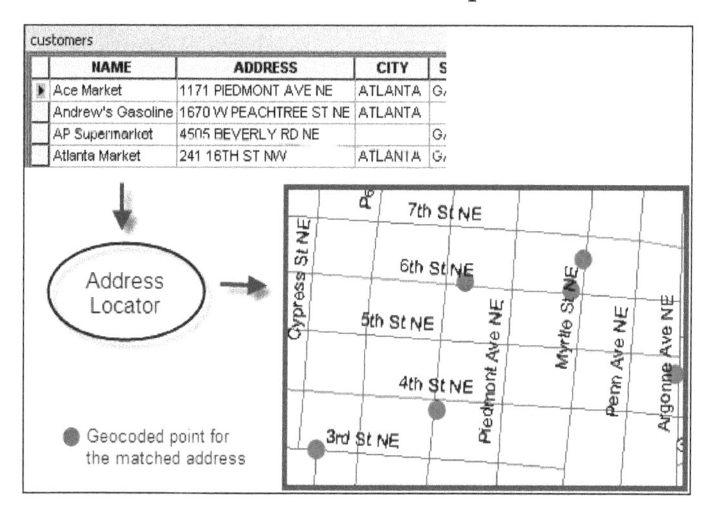

Usage

- You may geocode addresses that are stored in a single field or are split into multiple fields. A single input field stores the complete address, for example, 303 Peachtree St NE, Atlanta, GA 30308. Multiple fields are supported if the input addresses are split into multiple fields such as Address, City, State, and ZIP for a general United States address.
- The output feature class is saved in the same spatial reference as the address locator. Changing the spatial reference of the output feature class is possible by setting a different output coordinate system in the tool's environment settings.
- The output feature class, by default, stores a copy of input address and additional information such as score, status, and matched address of each record. The addresses can be rematched either using the Rematch Addresses tool or the Interactive Rematch dialog box

in ArcMap. Editing addresses in the input address table will not change the result in the output feature class once the matching process finishes and the feature class is created.

- Set the Dynamic Output Feature Class parameter to true (checked) if you want the matching result in the output feature class to be updated automatically when the input address table is updated. A relationship class is created for the input table and the output feature class. When an address in the input table is changed in an editing session in ArcMap, the address will be geocoded again immediately and the related record in the output feature class will be updated with the new geocoding result. The automatic update is also supported for adding a new record or deleting an existing record in the input table.
- An ArcGIS Online for organizations subscription is required to match a table of addresses using the ArcGIS Online World geocoding service.

Syntax

GeocodeAddresses_geocoding (in_table, address_locator, in_address_fields, out_feature_class, {out_relationship_type})

Parameter	Explanation	Data Type
in_table	The table of addresses to geocode.	Table View
address_locator	The address locator to use to geocode the table of addresses.	Address Locator
in_address_fields [input_address_ field, table_field_ name]	Each field mapping in this parameter is in the format input_address_ field, table_field_name where input_address_field is the name of the input address field specified by the address locator, and table_field_ name is the name of the corresponding field in the table of addresses you want to geocode. You may specify one single input field that stores the complete address. Alternatively, you may also specify multiple fields if the input addresses are split into different fields such as Address, City, State, and ZIP for a general United States address. If you choose not to map an optional input address field used by the address locator to a field in the input table of addresses, specify that there is no mapping by using <None> in place of a field name.	Field Info
out_feature_class	The output geocoded feature class or shapefile.	Feature Class
out_relationship_ type (Optional)	Indicates whether to create a static copy of the address table inside the geocoded feature class or to create a dynamically updated geocoded feature class. • STATIC —Creates a static copy of the fields input address table in the output feature class. This is the default. • DYNAMIC —Creates a relationship class between the input address table and output feature class so that edits to the addresses in the input address table are automatically updated in the output feature class. This option is supported only if the input address table and output feature class are in the same geodatabase workspace.	Boolean

Code Sample

```
# Import system modules

import arcpy

from arcpy import env

env.workspace = "C :\ArcTutor\Geocoding\atlanta.gdb"

# Set local variables:

address_table = "customers"

address_locator = "Atlanta_AddressLocator"

address_fields = "Street Address;City City;State State;ZIP Zip"

geocode_result = "geocode_result"

arcpy.GeocodeAddresses_geocoding(address_table, address_locator, address_fields, geo-
code_result, 'STATIC')
```

GEO URI SCHEME

The geo URI scheme is a Uniform Resource Identifier (URI) scheme defined by the Internet Engineering Task Force's RFC 5870 as:

A Uniform Resource Identifier (URI) for geographic locations using the 'geo' scheme name. A 'geo' URI identifies a physical location in a two- or three-dimensional coordinate reference system in a compact, simple, human-readable, and protocol-independent way.

The current revision of the vCard specification supports geo URIs in a vCard's "GEO" property, and the GeoSMS standard uses geo URIs for geotagging SMS messages. Android based devices support geo URIs, although that implementation is based on a draft revision of the specification, and supports a different set of URI parameters and query strings.

A geo URI is not to be confused with the site GeoUrl (which implements ICBM address).

A simple geo URI might look like:

```
geo:37.786971,-122.399677
```

where the two numerical values represent latitude and longitude respectively, and are separated by a comma. They are coordinates of a horizontal grid (2D). If a third comma-separated value is present, it represents altitude; so, coordinates of a 3D grid. Coordinates in the Southern and Western hemispheres as well as altitudes below the coordinate reference system (depths) are signed negative with a leading dash.

The geo URI also allows for an optional "uncertainty" value, separated by a semicolon, representing the uncertainty of the location in meters, and is described using the "u" URI parameter. A geo URI with an uncertainty parameter looks as follows:

```
geo:37.786971,-122.399677;u=35
```

A geo URI may, for example, be included on a web page, as HTML:

```
<a href="geo:37.786971,-122.399677;u=35">Wikimedia Headquarters</a>
```

so that a geo URI-aware user agent such as a web browser could launch the user's chosen mapping service; or it could be used in an Atom feed or other XML file.

Coordinate Reference Systems

The values of the coordinates only make sense when a coordinate reference system (CRS) is specified. The default CRS is the World Geodetic System 1984 (WGS-84), and it is not recommended to use any other:

The optional 'crs' URI parameter described below may be used by future specifications to define the use of CRSes other than WGS-84. This is primarily intended to cope with the case of another CRS replacing WGS-84 as the predominantly used one, rather than allowing the arbitrary use of thousands of CRSes for the URI (which would clearly affect interoperability).

The only justified use of other CRS today is, perhaps, to preserve projection in large-scale maps, as local UTM, or for non-terrestrial coordinates such as those on the Moon or Mars. The syntax and semantic of the CRS parameter, separated by a semicolon. Examples:

- The Washington Monument's location expressed with UTM-zone 18S and its standard ID:

  ```
  geo:323482,4306480;crs=EPSG:32718;u=20
  ```

- A geo URI for a hypothetical lunar CRS created in 2011 might be:

  ```
  geo:37.786971,-122.399677;crs=Moon-2011;u=35
  ```

The order in which the semicolon-separated parameters occur is partially significant. Whilst the labeltext parameter and future parameters may be given in any order, the crs and the u parameters must come first. If both are used, the crs must precede the u. All parameters are case-insensitive, so, imagining a future new parameter mapcolors, it can be ignored by simpler applications, and the above example is exactly equivalent to:

```
geo:323482,4306480;CRS=epsg:32718;U=20;mapcolors=for_daltonic
```

Being in doubt, remember that use of the lowercase representation of parameter names (crs u and mapcolors) is preferred.

Semantics and usual Interpretations

The Geo URI scheme semantics is not explicit about some mathematical assumptions, so it is open to interpretation. After ~10 years of its publication, there are some consensual or "most frequently used" assumptions.

Altitude

The syntax of the Geo UI defines coodinates as coordinates = coord-a "," coord-b ["," coord-c], where coord-c is optional. The semantic of coord-c for WGS-84 is altitude (specifically the "ground elevation", relative to the current geoid attached to WGS84), and the concept is extend for other coordinates (of non-default CRS).

The RFC explains that "undefined <altitude> MAY assume that the URI refers to the respective location on Earth's physical surface." However, "an <altitude> value of 0 MUST NOT be mistaken to refer to 'ground elevation'".

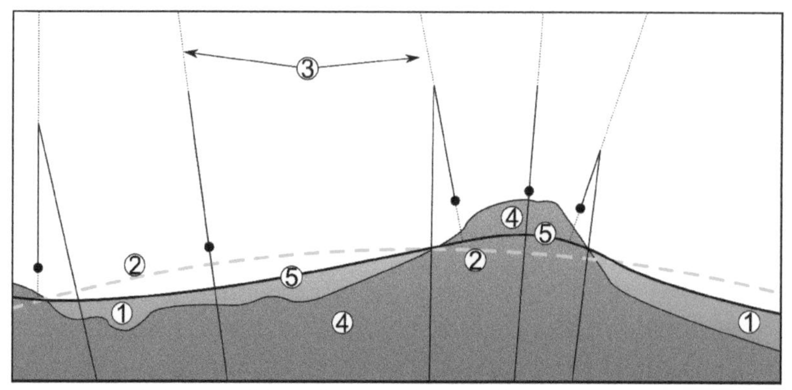

1. Ocean, 2. Reference ellipsoid, 3. Local plumb line, 4. Continent, 5. Geoid.

In other words, when an altitude is defined, the measurement is done relative to the geoid (#5; black line in the image), a surface defined by Earth's gravity approximating the mean sea level. When it is undefined, the elevation is assumed to be the altitude of the latitude-longitude point, that is its height (or negative depth) relative to the geoid (i.e. "ground elevation"). A point with a measure "altitude=0" is, however, not to be confused with an undefined value: it refers to an altitude of 0 meters above the geoid.

Uncertainty

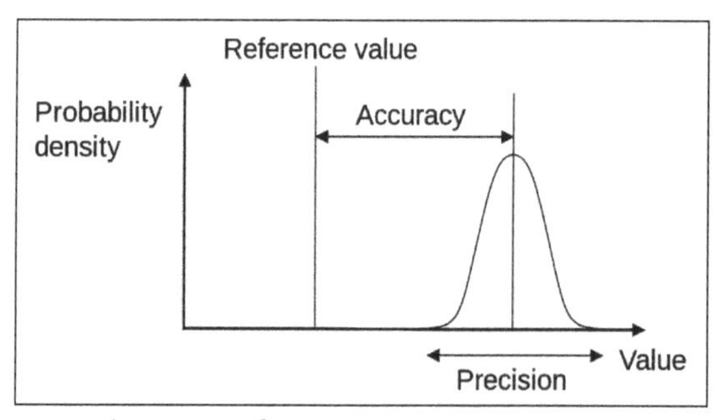

Facets of the uncertainty. According to ISO 5725-1: *accuracy* is the proximity of measurement results to the true value; *precision* is the degree to which repeated (or reproducible) measurements under unchanged conditions show the same results.

Geo URI is not about exact abstract positions, strictly it is an location estimate, and we can interpret it (from RFC 5870 and RFC 5491) as the approximate physical position of an object in the Earth's surface.

The RFC 5870 not formalize the use of the "*uncertainty*" term. The clues about it come from citations: the only normative reference with something about *uncertainty* is the RFC 5491. The main informative reference, ISO 6709:2008, not use the term "uncertainty", but use the terms "accuracy" and "precision", which are uncertainty facets and can be interpreted in accordance with ISO 5725-1 (illustrated).

Putting all together, adopting these clues, the usual statistical assumptions, and the explicit definitions of the RFC, we obtain the Geo URI's *uncertainty* mathematical properties:

- Uncertainty is symmetric: The RFC is explicit, and we can understand it as valid simplification hypothesis. Results in a spherical volume around the point (or a disk by 2D projection). By RFC 5491 "locations are expressed as a point and an area or volume of uncertainty around the point".
    - Using RFC 5491, we can suppose that "It is RECOMMENDED that uncertainty is expressed at a confidence of 95% or higher". Therefore, the uncertainty is two standard deviations, 2σ, and it is the radius of the disk that represents uncertainty geometrically.
- Fixed measure unit: The RFC obligate the use of meters as *uncertainty* measure units, even when coordinates (CRS) use other (like default that is decimal degrees). It is a semantic and a convertion problem: the
- Gaussian error model: RFC say nothing, we interpreting the phrases "amount of uncertainty in the location" and "the uncertainty with which the identified location of the subject is known", all in the context of the normative reference, RFC 5491 (and the informative references like ISO 6709:2008).
    - Adopting standard error model: The model of the most common descriptive statistics modeling.
    - It is imposed, is independent of selection process of *uncertainty* description, there are no other choices.
- Total uncertainty: It Is only one parameter representing "all uncertainty", the uncertainty in the spatial measure and uncertainty about object definition or object's center. It is a sum of random variables. There is no simplification hypothesis defined to reduce it to a one-variable model.

Imagining the location of an ant colony to illustrate:

- The colony is a 3D object at the (exactly) the Terrain surface, so at precise altitude (approximated to a zero uncertainty measure).
- The 3D object has some consensual definition, but it is not precise, so, its uncertainty can not be neglected. This lack of precision can be about the fact that the anthill is hidden under the ground (it is an "estimated object"), or the formal definition of its delimitation, etc. This kind of uncertainty has no correlation with the location (e.g. GPS) uncertainty measure.
    - The disk representing the anthill (as uncertainty of the object) is modeled as 2σ to be a 95% of confidence area.
- The point is a GPS location measure, that is, the "center" of the projection of the 3D object in the 2D surface.

The total uncertainty is the sum of GPS error and object-definition error. The latitude and longitude GPS errors need to be simplified (to a disk) and converted into meters. If the errors were inferred from a different model, they need to be converted to the Gaussian model.

Unofficial Extensions

Some vendors, such as Android OS, have adopted extensions to the "geo" URI scheme:

- z: Zoom level for Web Mercator projection scaling. The value is an integer from 1 to 21.
- q: Perform a search for the keyword given around the point. If the location is given as "0,0", search around the current position. A parenthetical can be used to indicate the label to show on the map.

Android adopts an unconventional approach to parsing the points: it does not show a map pin at the point given normally, and a map pin will show up only when given as the query. In other words, to show a pin at the Wikimedia Foundation office, one should not use geo:37.78918,-122.40335 but geo:0,0?q=37.78918,-122.40335.

GEOHASH

Geohash is a public domain geocode system invented in 2008 by Gustavo Niemeyer, which encodes a geographic location into a short string of letters and digits. It is a hierarchical spatial data structure which subdivides space into buckets of grid shape, which is one of the many applications of what is known as a Z-order curve, and generally space-filling curves.

Geohashes offer properties like arbitrary precision and the possibility of gradually removing characters from the end of the code to reduce its size (and gradually lose precision). As a consequence of the gradual precision degradation, nearby places will often (but not always) present similar prefixes. The longer a shared prefix is, the closer the two places are.

Web Site Geohash.Org

In February 2008, together with the announcement of the system, the inventor launched the website http://geohash.org/, which allows users to convert geographic coordinates to short URLs which uniquely identify positions on the Earth, so that referencing them in emails, forums, and websites is more convenient.

To obtain the Geohash, the user provides an address to be geocoded, or latitude and longitude coordinates, in a single input box (most commonly used formats for latitude and longitude pairs are accepted), and performs the request.

Besides showing the latitude and longitude corresponding to the given Geohash, users who navigate to a Geohash at geohash.org are also presented with an embedded map, and may download a GPX file, or transfer the waypoint directly to certain GPS receivers. Links are also provided to external sites that may provide further details around the specified location.

For example, the coordinate pair 57.64911,10.40744 (near the tip of the peninsula of Jutland, Denmark) produces a slightly shorter hash of u4pruydqqvj.

Uses

The main usages of Geohashes are:

- As a unique identifier.
- To represent point data, e.g. in databases.

Geohashes have also been proposed to be used for geotagging.

When used in a database, the structure of geohashed data has two advantages. First, data indexed by geohash will have all points for a given rectangular area in contiguous slices (the number of slices depends on the precision required and the presence of geohash "fault lines"). This is especially useful in database systems where queries on a single index are much easier or faster than multiple-index queries. Second, this index structure can be used for a quick-and-dirty proximity search: the closest points are often among the closest geohashes.

Algorithm and Example

Using the hash ezs42 as an example, here is how it is decoded into a decimal latitude and longitude.

The first step is decoding it from a variant of base 32 using all digits 0-9 and almost all lower case letters except a, i, l and o like this:

Decimal	0	1	2	3	4	5	6	7	8	9	10	11	12	13	14	15
Base 32	0	1	2	3	4	5	6	7	8	9	b	c	d	e	f	g
Decimal	16	17	18	19	20	21	22	23	24	25	26	27	28	29	30	31
Base 32	h	j	k	m	n	p	q	r	s	t	u	v	w	x	y	z

This operation results in the bits 01101 11111 11000 00100 00010. Assuming that counting starts at 0 in the left side, the even bits are taken for the longitude code (0111110000000), while the odd bits are taken for the latitude code (101111001001).

Each binary code is then used in a series of divisions, considering one bit at a time, again from the left to the right side. For the latitude value, the interval -90 to +90 is divided by 2, producing two intervals: -90 to 0, and 0 to +90. Since the first bit is 1, the higher interval is chosen, and becomes the current interval. The procedure is repeated for all bits in the code. Finally, the latitude value is the center of the resulting interval. Longitudes are processed in an equivalent way, keeping in mind that the initial interval is -180 to +180.

For example, in the latitude code 101111001001, the first bit is 1, so we know our latitude is somewhere between 0 and 90. Without any more bits, we'd guess the latitude was 45, giving us an error of ±45. Since more bits are available, we can continue with the next bit, and each subsequent bit halves this error. This table shows the effect of each bit. At each stage, the relevant half of the range is highlighted in green; a low bit selects the lower range, a high bit selects the upper range.

The column "mean value" shows the latitude, simply the mean value of the range. Each subsequent bit makes this value more precise.

Latitude code 101111001001						
bit position	bit value	min	mid	max	mean value	maximum error
0	1	-90.000	0.000	90.000	45.000	45.000
1	0	0.000	45.000	90.000	22.500	22.500
2	1	0.000	22.500	45.000	33.750	11.250
3	1	22.500	33.750	45.000	39.375	5.625
4	1	33.750	39.375	45.000	42.188	2.813
5	1	39.375	42.188	45.000	43.594	1.406
6	0	42.188	43.594	45.000	42.891	0.703
7	0	42.188	42.891	43.594	42.539	0.352
8	1	42.188	42.539	42.891	42.715	0.176
9	0	42.539	42.715	42.891	42.627	0.088
10	0	42.539	42.627	42.715	42.583	0.044
11	1	42.539	42.583	42.627	42.605	0.022

Longitude code 0111110000000						
bit position	bit value	min	mid	max	mean value	maximum error
0	0	-180.000	0.000	180.000	-90.000	90.000
1	1	-180.000	-90.000	0.000	-45.000	45.000
2	1	-90.000	-45.000	0.000	-22.500	22.500
3	1	-45.000	-22.500	0.000	-11.250	11.250
4	1	-22.500	-11.250	0.000	-5.625	5.625
5	1	-11.250	-5.625	0.000	-2.813	2.813
6	0	-5.625	-2.813	0.000	-4.219	1.406
7	0	-5.625	-4.219	-2.813	-4.922	0.703
8	0	-5.625	-4.922	-4.219	-5.273	0.352
9	0	-5.625	-5.273	-4.922	-5.449	0.176
10	0	-5.625	-5.449	-5.273	-5.537	0.088
11	0	-5.625	-5.537	-5.449	-5.581	0.044
12	0	-5.625	-5.581	-5.537	-5.603	0.022

Final rounding should be done carefully in a way that

$$\text{Min} \le \text{round (value)} \le \text{max}$$

So while rounding 42.605 to 42.61 or 42.6 is correct, rounding to 43 is not.

Number of geohash characters and precision in km

geohash length	lat bits	lng bits	lat error	lng error	km error
1	2	3	±23	±23	±2500
2	5	5	±2.8	±5.6	±630
3	7	8	±0.70	±0.70	±78

4	10	10	±0.087	±0.18	±20
5	12	13	±0.022	±0.022	±2.4
6	15	15	±0.0027	±0.0055	±0.61
7	17	18	±0.00068	±0.00068	±0.076
8	20	20	±0.000085	±0.00017	±0.019

Limitations when used for Deciding Proximity

Edge Cases

Geohashes can be used to find points in proximity to each other based on a common prefix. However, edge case locations close to each other but on opposite sides of the 180 degree meridian will result in Geohash codes with no common prefix (different longitudes for near physical locations). Points close by at the North and South poles will have very different geohashes (different longitudes for near physical locations).

Two close locations on either side of the Equator (or Greenwich meridian) will not have a long common prefix since they belong to different 'halves' of the world. Put simply, one location's binary latitude (or longitude) will be 011111... and the other 100000...., so they will not have a common prefix and most bits will be flipped. This can also be seen as a consequence of relying on the Z-order curve (which could more appropriately be called an N-order visit in this case) for ordering the points, as two points close-by might be visited at very different times. However, two points with a long common prefix will be close-by.

In order to do a proximity search, one could compute the southwest corner (low geohash with low latitude and longitude) and northeast corner (high geohash with high latitude and longitude) of a bounding box and search for geohashes between those two. This search will retrieve all points in the z-order curve between the two corners, which can be far too many points. This method also breaks down at the 180 meridians and the poles. Solr uses a filter list of prefixes, by computing the prefixes of the nearest squares close to the geohash.

Non-linearity

Since a geohash (in this implementation) is based on coordinates of longitude and latitude the distance between two geohashes reflects the distance in latitude/longitude coordinates between two points, which does not translate to actual distance,

Example of non-linearity for latitude-longitude system:

- At the Equator (0 Degrees) the length of a degree of longitude is 111.320 km, while a degree of latitude measures 110.574 km, an error of 0.67%.
- At 30 Degrees (Mid Latitudes) the error is 110.852/96.486 = 14.89%.
- At 60 Degrees (High Arctic) the error is 111.412/55.800 = 99.67%, reaching infinity at the poles.

Note that these limitations are not due to geohashing, and not due to latitude-longitude coordinates, but due to the difficulty of mapping coordinates on a sphere (non linear and with wrapping of values, similar to modulo arithmetic) to two dimensional coordinates and the difficulty of exploring a two dimensional space uniformly. The first is related to Geographical coordinate system and Map projection, and the other to Hilbert curve and z-order curve. Once a coordinate system is found that represents points linearly in distance and wraps up at the edges, and can be explored uniformly, applying geohashing to those coordinates will not suffer from the limitations above.

While it is possible to apply geohashing to an area with a cartesian coordinate system, it would then only apply to the area where the coordinate system applies.

Despite those issues, there are possible workarounds, and the algorithm has been successfully used in Elasticsearch, MongoDB, HBase, and Accumulo to implement proximity searches.

An alternative to storing Geohashes as strings in a database are Locational codes, which are also called spatial keys and similar to QuadTiles.

GEOCODING

Geocoding is the process of transforming a description of a location—such as a pair of coordinates, an address, or a name of a place—to a location on the earth's surface. You can geocode by entering one location description at a time or by providing many of them at once in a table. The resulting locations are output as geographic features with attributes, which can be used for mapping or spatial analysis.

You can quickly find various kinds of locations through geocoding. The types of locations that you can search for include points of interest or names from a gazetteer, like mountains, bridges, and stores; coordinates based on latitude and longitude or other reference systems, such as the Military Grid Reference System (MGRS) or the U.S. National Grid system; and addresses, which can come in a variety of styles and formats, including street intersections, house numbers with street names, and postal codes.

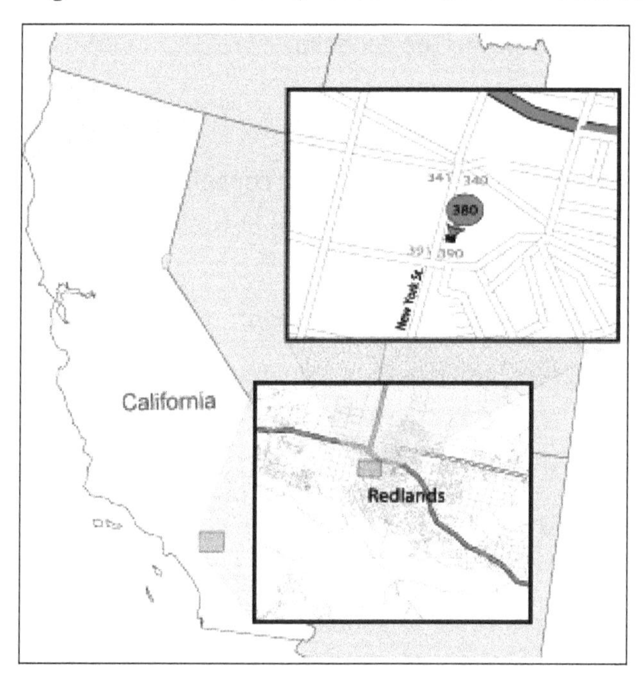

What can Geocoding be used for?

From simple data analysis to business and customer management to distribution techniques, there is a wide range of applications for which geocoding can be used. With geocoded addresses, you can spatially display the address locations and recognize patterns within the information. This can be done by simply looking at the information or using some of the analysis tools available with ArcGIS. One can also display your address information based on certain parameters, allowing you to further analyze the information.

Address Data Analysis

With geocoded addresses, you can spatially display the address locations and begin to recognize patterns within the information. This can be done by simply looking at the information or by using some of the analysis tools available with ArcGIS. One can also display your address information based on certain parameters, allowing you to further analyze the information.

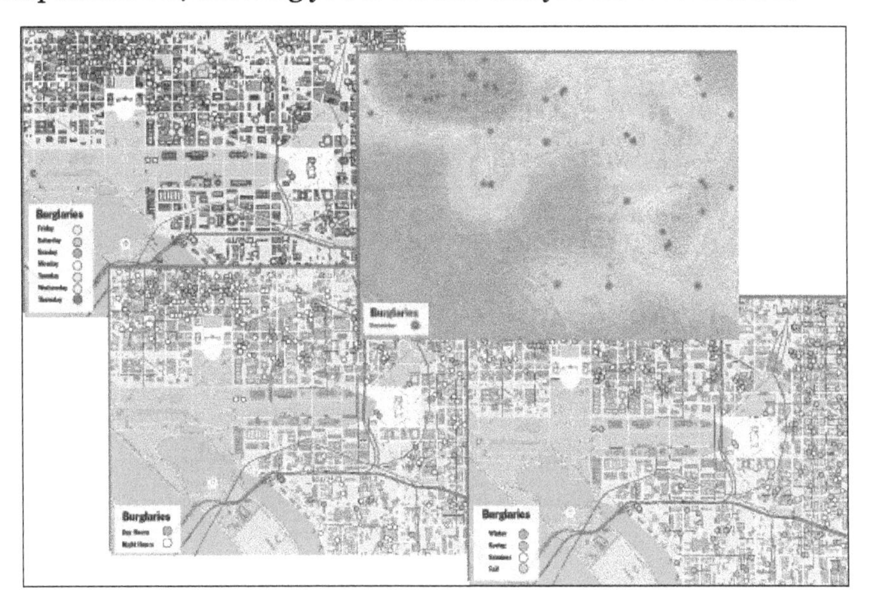

The annual record of burglaries was initially created by geocoding a database table of burglaries that consisted of an address for each. The screen shots above show how the geocoded addresses were presented according to time, season, and day of the week to assist in crime prevention planning. Additional analysis tools available in ArcGIS could be used to further analyze the information to identify patterns. Base data for this sample application was provided by Tele Atlas and Vexcel Corp.

Customer Data Management

Geocoding acts as a crucial part of customer data management. Nearly every organization maintains address information for each customer or client. This is usually in tabular format, containing the customer name, address, buying habits, and any other information you have collected. Geocoding allows you to take your customers' information and create a map of their locations. Using a variety of related applications, you can use this information in many ways, from establishing marketing strategies to targeting specific clusters of customers to producing route maps and directions. The geocoded locations of your customers can be invaluable data.

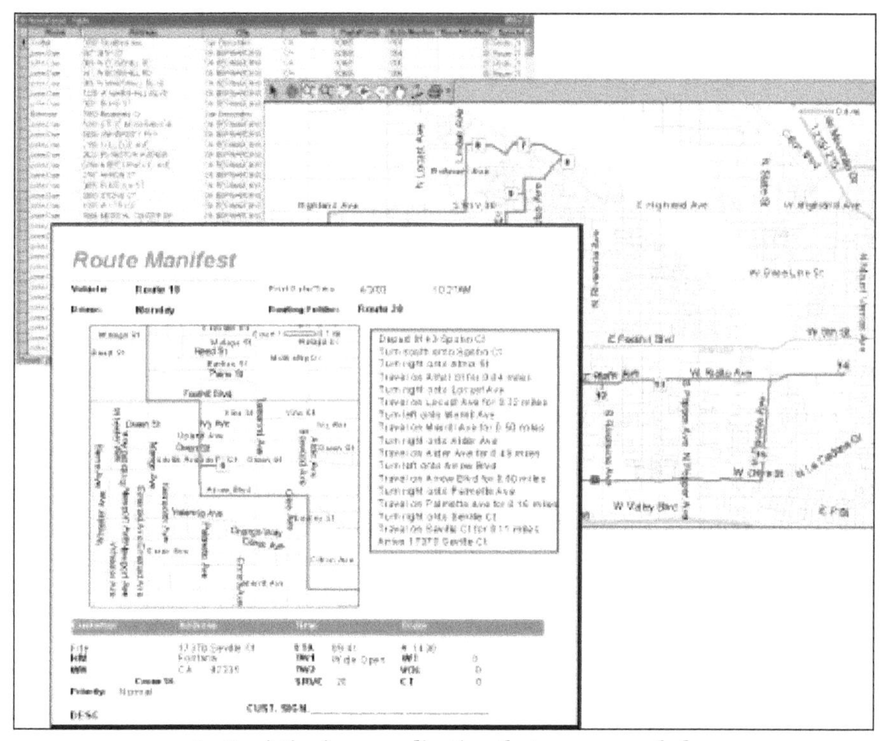

ArcLogistics is an application that uses geocoded addresses to optimally plan customer deliveries.

Distributed Geocoding Applications

You can use various methods to share your geocoding functionality. These include everything from collecting the address locators and sharing via a compressed file or compact disc to developing an online application, allowing users to do geocoding over the Internet.

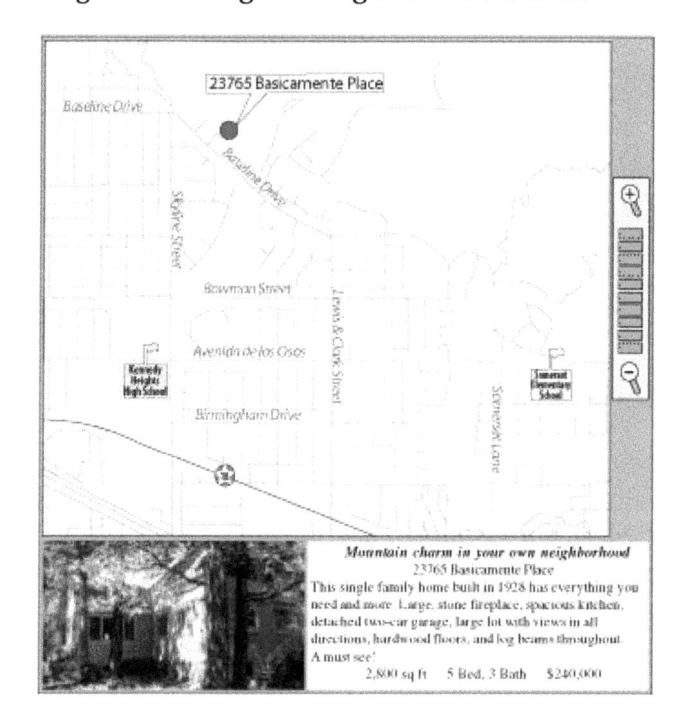

Many real estate firms have found advantages in distributing information about available real estate via the Internet. By combining the database of available homes and ArcGIS Web services, you can distribute the spatial and nonspatial information about a home to a wide audience.

Geocoding using ArcGIS

In the real world, you find locations based on some description. This might be a number and street name. It might include the name of the city, state, or country or natural features, such as a drainage basin or ecological region.

For example, if you needed to locate the address 380 New York St., Redlands, CA 92373 with the right street map, it would not take you long to find the exact location. You might first find California, then find the city of Redlands. You might also use a postal code map and locate the region covered by the corresponding ZIP Code value. You would then locate the street and interpret where and on which side of the 300 block the address is located.

Just as you found the address by narrowing your search to a specific region, finding a particular feature, and interpreting a point along that feature, the computer is doing the same process to assign a location to an address when geocoding. Geocoding starts with a textual description of a location and translates that into the x,y coordinates that can be plotted on a map.

The Value of Good Reference Data

Your first step when you want to find something on a map is to have the right map. There is no way you will find your way to 380 New York Street in Redlands, California, if you only have a map of Canada. Also, you won't be able to pinpoint the address very well if your map only shows highways and major cities. Your map must have enough detail of the area to pinpoint the location for which you are searching.

It is no different when geocoding in ArcGIS. The layers that you use for creating an address locator, known as reference data, need to have details of the specific point you want to find. When looking for addresses, the primary reference data usually consists of a street network, but a parcel map can be used as well. The important thing is that the data has the detail that you want to find.

Address Locator

The address locator is the major component in the geocoding process. An address locator is created based on a specific address locator style. Once created, an address locator contains the geocoding properties and parameters that are set on the Address Locator Properties dialog box, a snapshot of the address attributes in the reference data, and the queries for performing a geocoding search. The address locator also contains a set of address parsing and matching rules that directs the geocoding engine to perform address standardization and matching.

What the Address Locator does?

Think of the address locator as a street guide or map book that you use to look up an address; it directs you to the page and pinpoints the location of the address. When you enter an address you want to find, the geocoding engine converts the input address into pieces, such as number, street name, and street type, based on the parsing rules defined in the address locator. These pieces are known as address elements. The geocoding engine may generate multiple interpretations of the same address, as some values in the input address can be considered in more than one element. For example, the word park can be both a street name and a street type. Each combination of the address elements will be searched in the address locator. The goal is to find all the possible matching candidates. Once possible candidates are identified, each individual variable in the candidate is compared with each corresponding address element. A score is generated indicating how well the address is matched. Finally, the address locator presents the best matches based on the score and the location of the address being matched.

GEOLOCATION

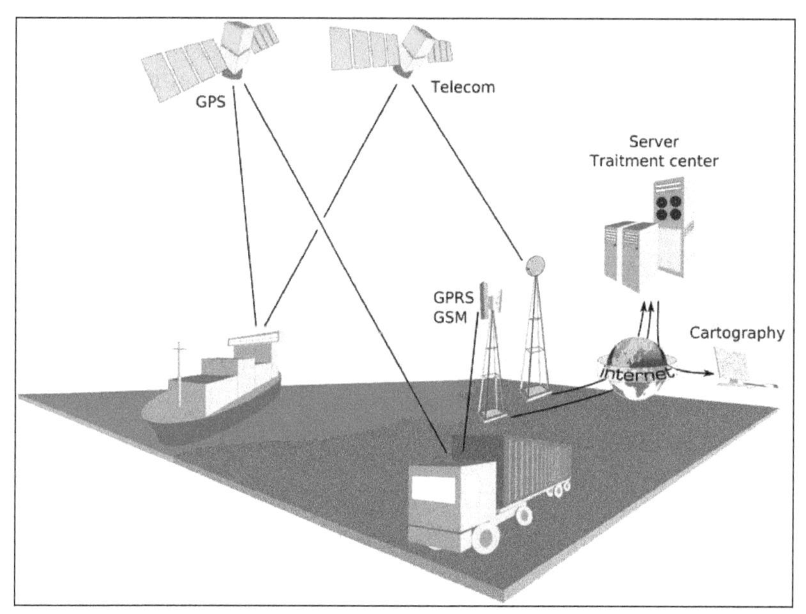

Principles of geolocation using GPS.

Geolocation is the identification or estimation of the real-world geographic location of an object, such as a radar source, mobile phone, or Internet-connected computer terminal. In its simplest form geolocation, involves the generation of a set of geographic coordinates and is closely related to the use of positioning systems, but its usefulness is enhanced by the use of these coordinates to determine a meaningful location, such as a street address.

The word *geolocation* also refers to the latitude and longitude coordinates of a particular location. The term and definition have been standardized by real-time locating system standard ISO/IEC 19762-5:2008.

In the field of animal biology and ecology, the word *geolocation* is also used to refer to the process of inferring the location of a tracked animal based, for instance, on the time history of sunlight brightness or the water temperature and depth measured by an instrument attached to the animal. Such instruments are commonly called archival tags (including microchip implants, Pop-up satellite archival tags, and data storage tags) or dataloggers.

Techniques

For either geolocating or positioning, the locating engine often uses radio frequency (RF) location methods, for example Time Difference Of Arrival (TDOA) for precision. TDOA systems often use mapping displays or other geographic information system. When satellite navigation (such as GPS) signals are unavailable, geolocation applications can use information from cell towers to triangulate the approximate position, a method that is not as accurate as GPS but has greatly improved in recent years. This is in contrast to earlier radiolocation technologies, for example Direction Finding where a line of bearing to a transmitter is achieved as part of the process.

Internet and computer geolocation can be performed by associating a geographic location with the Internet Protocol (IP) address, MAC address, RFID, hardware embedded article/production number, embedded software number (such as UUID, Exif/IPTC/XMP or modern steganography), invoice, Wi-Fi positioning system, device fingerprint, canvas fingerprinting or device GPS coordinates, or other, perhaps self-disclosed information.

IP address location data can include information such as country, region, city, postal/zip code, latitude, longitude and time zone. Deeper data sets can determine other parameters such as domain name, connection speed, ISP, language, proxies, company name, US DMA/MSA, NAICS codes, and home/business.

At times geolocation can be more deductive, as with crowdsourcing efforts to determine the position of videos of training camps, combats, and beheadings in Syria by comparing features detected in the video with publicly available map databases such as Google Earth.

TOPONYM RESOLUTION

Toponym resolution is the relationship process between a toponym, i.e. the mention of a place, and an unambiguous spatial footprint of the same place.

To map a set of place names or toponyms that occur in a document to their corresponding latitude/longitude coordinates, a polygon, or any other spatial footprint, a disambiguation step is necessary. A toponym resolution algorithm is an automatic method that performs a mapping from a toponym to a spatial footprint.

Most methods for toponym resolution employ a gazetteer of possible mappings between names and spatial footprints.

Resolution Process

The “unambiguous spatial footprint of the same place” of definition can be in fact unambiguous, or “not so unambiguous”. There are some different *contexts of uncertainty* where the resolution process can occur:

- When the evidence is geographical and with no uncertainty. For example, to obtain the country name of a photo place, when the place is a GPS position (10 meters of error), at 1000 km far from country borders.

- When the evidence is geographical, but with considerable uncertainty. Imagine a similar scenario where the GPS error is 100 meters and the place is near from, ~100 meters, of the country borders.
- When the evidence is only textual. Imagine a letter where the narrator is a tourist telling about his trip after he returned from vacation. The only evidences are textual, in the narrative.
- Mixed sources of evidence: more than one evidence, no one precise.

From Geographical Evidence

The toponym resolution sometimes is a simple conversion from name to abbreviation, in special when the abbreviation is used as standard geocode. For example converting the official country name Afghanistan into an ISO contry code, AF.

In annotating media and metadata, the conversion using a map and the geographical evidence (e.g. GPS), is the most usual approach to obtain toponym, or a geocode that represents the toponym.

Approaches

Toponym resolution methods can be generally divided into supervised and unsupervised models. Supervised methods typically cast the problem as a learning task wherein the model first extracts contextual and non-contextual features and then, a classifier is trained on a labelled dataset. Adaptive model is one of the prominent models proposed in resolving toponyms. For each interpretation of a toponym, the model derives context-sensitive features based on geographical proximity and sibling relationships with other interpretations. In addition to context related features, the model benefits from context-free features including population, and audience location. On the other hand, unsupervised models do not warrant annotated data. They are superior to supervised models when the annotated corpus is not sufficiently large, and supervised models may not generalize well.

Unsupervised models tend to better exploit the interplay of toponyms mentioned in a document. The Context-Hierarchy Fusion model estimates the geographic scope of documents and leverages the connections between nearby place names as evidence to resolve toponyms. By means of mapping the problem to a conflict-free set cover problem, this model achieves a coherent and robust resolution.

Furthermore, adopting Wikipedia and knowledge bases have been shown effective in toponym resolution. TopoCluster models the geographical senses of words by incorporating Wikipedia pages of locations and disambiguates toponyms using the spatial senses of the words in the text.

Geoparsing

Geoparsing is a special toponym resolution process of converting free-text descriptions of places (such as "twenty miles northeast of Jalalabad") into unambiguous geographic identifiers, such as geographic coordinates expressed as latitude-longitude. One can also geoparse location references from other forms of media, for examples audio content in which a speaker mentions a place. With geographic coordinates the features can be mapped and entered into Geographic Information Systems. Two primary uses of the geographic coordinates derived from unstructured content are to plot portions of the content on maps and to search the content using a map as a filter.

Geoparsing goes beyond geocoding. Geocoding analyzes unambiguous structured location references, such as postal addresses and rigorously formatted numerical coordinates. Geoparsing handles ambiguous references in unstructured discourse, such as "Al Hamra," which is the name of several places, including towns in both Syria and Yemen.

A geoparser is a piece of software or a (web) service that helps in this process. Some examples:

- GEOLocate automated georeferencing.
- BioGeomancer: Semi-automatic georeferencing.
- GEOnet Names Server: Freely available GIS information for areas outside of the U.S.A. and Antarctica, updated monthly by the National Geospatial-Intelligence Agency (NGA) and the U.S. Board on Geographic Names (US BGN).
- Geographic Names Information System (GNIS): Freely available database containing information on almost 2 million physical features, places, and landmarks in the U.S.A.
- CLAVIN: CLAVIN (Cartographic Location And Vicinity INdexer) is an open source software package for document geotagging and geoparsing that employs context-based geographic entity resolution.
- Geoparser.io: Geoparser.io is a web service that identifies places mentioned in text, disambiguates those places, and returns GeoJSON with detailed metadata about the places found in the text.
- Geocode.xyz: Geocode.xyz is a web service that identifies both place names and street addresses mentioned in text.
- Geoparsepy: Geoparsepy is a free Python geoparsing library supporting free text location identification and disambiguation using the OpenStreetMap database.

REVERSE GEOCODING

"Reverse Geocoding" is the process of finding an address, toponym or an other type of resource for a given lat/lng pair. GeoNames offers a wide range of reverse geocoding web services.

- Find nearby postal codes : finds postal codes and place names for the given lat/lng within a given radius. (supported countries).
- Find nearby place name: Finds pouplated place names for the given lat/lng within a given radius. Simlar to findNearby, but with filter for populated places.
- Find nearby: Finds any toponym for the given lat/lng within a given radius. Similar to findNearbyPlaceName, but without the filter for populated places.
- Extended Find nearby: Most detailed information for toponym for the given lat/lng within a given radius (combination of 4 services)
- Country: Finds the ISO country code for the given lat/lng.

- Country Subdivision: Finds the country and the administrative subdivison for the given lat/lng.
- Ocean: Finds the ocean or sea for the given lat/lng.
- Neighbourhood: Finds the neighbourhood for the given lat/lng (US cities only).
- Weatherstation and weather observation : finds weatherstations and most recent weather observations for the given lat/lng.

Address reverse geocoding (selected countries):

- Address: Finds the nearest address for a lat/lng.

US Street level reverse geocoding services:

- Find nearest Address: Finds the nearest street and address for a given lat/lng pair. (US only).
- Find nearest Intersection: Finds the nearest street and the next crossing street for a given lat/lng pair. (US only).
- Find nearby Streets: Finds the nearest street segments for a given lat/lng pair. (US only).

Global Street level reverse geocoding services (cc by-sa license):

- Find nearest Intersection: Finds the nearest street and the next crossing street for a given lat/lng pair.
- Find nearby Streets: Finds the nearest street segments for a given lat/lng pair.

Others:

- Timezone: Returns the timezone at the given lat/lng.
- Aster Elevation:Returns the Aster elevation (30mx30m).
- SRTM3 Elevation: Returns the SRTM3 elevation (90mx90m).
- GTOPO30 Elevation: Return the GTOPO30 elevation (1Kmx1Km).
- Points of Interests: Returns nearby points of interests (cc by-sa license).

Reverse geocodes create addresses from point locations in a feature class. The reverse geocoding process searches for the nearest address or intersection for the point location based on the specified search distance.

Usage

- The input feature class must contain point shapes with valid XY coordinates. Addresses will not be returned on points with null coordinates.
- The output feature class will contain the same number of records as in the input feature class. Additional fields containing the result addresses are added to the feature class. The names of the fields are prefixed with REV_. If an address cannot be found, the fields will contain empty values.

- If the spatial reference of the input feature class is different from the address locator, the address locator will transform the coordinates on the fly and try to find the match. The output feature class will be saved in the same spatial reference as the input feature class. Changing the spatial reference of the output feature class is possible by setting a different output coordinate system in the tool's environment settings.
- If a point in the input feature class fails to return an address, it means there are no features in the address locator that can be associated with the input point. Here are a few common causes for the unmatched points:
    - The search distance is too small that the point cannot find any nearest features.
    - The point contains null coordinates.
    - The point's coordinates are incorrect and cannot be transformed to the spatial reference used in the address locator.
    - The address locator does not contain reference features in the area that can be associated with the point.

You may increase the search distance so that the chance to find the nearest address is higher, or use a different address locator that contains more features or covers a bigger area to match the input points.

- An ArcGIS Online for organizations subscription is required if you reverse geocode a feature class using the ArcGIS Online Geocoding Service.

Syntax

ReverseGeocode_geocoding (in_features, in_address_locator, out_feature_class, {address_type}, {search_distance}).

Parameter	Explanation	Data Type
in_features	A point feature class or layer from which addresses are returned based on the features' point location.	Feature Class
in_address_locator	The address locator to use to reverse geocode the input feature class.	Address Locator
out_feature_class	The output feature class.	Feature Class
address_type (Optional)	Indicates whether to return addresses for the points as street addresses or intersection addresses if the address locator supports intersection matching. • ADDRESS —Returns street addresses or in the format defined by the input address locator. This is the default option. • INTERSECTION —Returns intersection addresses. This option is available if the address locator supports matching intersection addresses.	String
search_distance (Optional)	The distance used to search for the nearest address or intersection for the point location.	Linear unit

Code Sample

ReverseGeocode Example (Python Window)

The following Python window script demonstrates how to use the ReverseGeocode function in immediate mode.

```
# Import system modules

import arcpy

from arcpy import env

env.workspace = "C:/data/locations.gdb"

# Set local variables:

input_feature_class = "customers"

address_locator = "e:/StreetMap/data/Street_Addresses_US"

result_feature_class = "customers_with_address"

arcpy.ReverseGeocode_geocoding(input_feature_class,  address_locator,  result_
feature_class, "ADDRESS", "100 Meters")
```

References

- What-are-geocodes-and-why-are-they-important-2, news: runneredq.com, Retrieved 19 April, 2019
- Kamalloo, ehsan; rafiei, davood (2018). A coherent unsupervised model for toponym resolution. Proceedings of the 2018 world wide web conference. Pp. 1287–1296. Arxiv:1805.01952. Doi:10.1145/3178876.3186027
- Geocode-addresses, geocoding-toolbox, tools, arcmap: arcgis.com, Retrieved 12 July, 2019 "rfc 5870 - a uniform resource identifier for geographic locations (geo uri)". Internet engineering task force. 2010-06-08. Retrieved 9 june 2010
- What-is-geocoding, geocoding, manage-data, latest, arcmap: arcgis.com, Retrieved 16 January, 2019
- Galuardi, benjamin; lam, chi hin (tim) (2014). "telemetry analysis of highly migratory species". Stock identification methods. Pp. 447–476. Doi:10.1016/b978-0-12-397003-9.00019-9. Isbn 9780123970039
- Reverse-geocode, geocoding-toolbox, tools, arcmap: arcgis.com, Retrieved 15 July, 2019
- crowd-funded journalists geo-locate isis training camp using the militants' own photos". Petapixel. 2014-08-25. Archived from the original on 2014-09-24. Retrieved 2014-09-20
- Reverse-geocoding, export: geonames.org, Retrieved 29 March, 2019

6

Applications of Geographic Information System

The geographic information systems are applied in various areas. Some of these include industries, archaeology, public health, aquatic science, environment, Google maps, Google Earth, geospatial intelligence, etc. This chapter discusses in detail these diverse applications of geographic information system.

GIS APPLICATIONS IN INDUSTRIES

Telecom Sector GIS Uses

Telecom sector is one of the vast sector which requires to get the details of all the regions to provide services to the people in each corner of the country. Telecom sector faces many problems which can be easily solved with the help of GIS.

The problems which are faced by Telecom sectors are :

- Capacity Planning: The main problem for the telecom sector is the planning of capacity. Capacity here refers to the capacity of network providing, which are the areas where we need to establish more network services. So for establishing network services in those areas they need some resources

So all these things can be easily found out with the help of GIS maps.

- Market Segmentation: Market segmentation states that its a concept to understand the customers and the use of the product by these customers, and specially to understand the market that needs to be target.
- Real – time knowledge of Network Structure : Telecom service providers can easily access to the points from where they provide the network service and how strong or weak the signal is in which area. GIS helps them to analyze the Network structure and status as well.
- optimum Use of Resources: GIS system helps the companies to utilize their resources efficiently and effectively, as GIS gives the information about where what is required, which helps these companies to invest where it is required. This leads to the efficient use of resources.

- Demand Forecasting : Once the company knows about the target market or target customers, it makes so easy for them to forecast the demand.

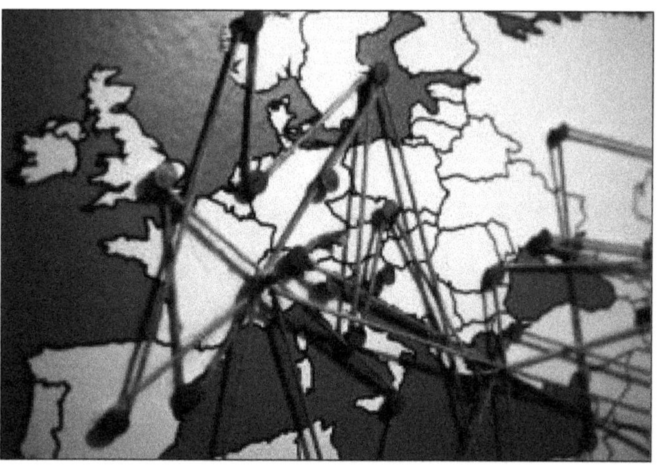

Telecom – GIS uses and application.

Hospital Industry GIS uses and Application

Hospital industry is the need of today's era, this is the only industry which is helping us to come out of any severe disease or it helps us whenever we need, they do not see time to serve us. We reach to them in midnight and they will be ready to serve you. So GIS is helping Hospital industry to serve us better in the following ways:

- Identifying health Trends: GIS helps this industry to identify and analyze the health trends, like which are the areas that can get affected with any particular disease,which are the areas where cancer patients are more and so on.
- Tracking the disease spread: GIS helps this industry to look over the spread a disease can cause. How much harmful a disease spread would be.

Hospital – GIS uses and application.

- Improved Services : GIS helps this industry to grow by providing best services. The services of this industry can be improved on a regular basis with the help of GIS. As GIS would let them know about which are the disease prone area, where what kind of disease can occur.

Agriculture GIS Application

Agriculture is the main source of income in many families, with Agriculture we get our two meals for a day, and it is very essential sector for any economy. We use agricultural products regularly,but have you ever seen a question crossing your mind that how does the agricultural products are produced ? So, there are few factors which can effect the Agriculture tremendously if not considered correct way. They are :

- Know Land Type: GIS helps you in looking for the type of land whether it is fertile or non fertile. Fertile land is used for agriculture purpose while non fertile is not usable or agriculture purpose.
- Weather Prone Area: GIS helps to locate the weather prone areas which can help the people to plan their agriculture process accordingly.
- Drought : GIS helps people to identify the area facing drought such that they can provide the crops to these areas.
- Water supply: water plays a major role in planning for agriculture as to we need water for irrigating the crops and that water should be poured to it in the right amount and it should be absorbed by the soil instead of leaking out of farms.

Traffic GIS uses and Application

Traffic is such an irresistible thing which has power to irritate us, to frustrate us and what not?? How great would that be to know the live traffic while leaving from home or office such that you can avoid that congestion?

Know live Congestion or future traffic on Google map.

You may also check how Google Map can be used for knowing real time traffic and predicting in future.

GIS can help you out with this problem as it can provide you the traffic updates using Real – Time data. Moreover, you can also predict the future traffic situations with the help of GIS. What else do you want?. So avoid the traffic and reach anywhere on time.

Economic Development GIS uses and Application

Economic development can be done using GIS. Basically, Economic development refers to the growth of economies of a country, which can be majorly effected by the business. So, Economic development is looking for the people to plan a business and invest money in it. But the question arises is how GIS can help in economic development?

The answer to this question is that GIS helps in identifying the locations and areas with all the attributes which can make a business grow with a fast or steady pace, GIS will find the best location for you, where you can plan a business and it will help you out with the business growth opportunities as well.

Insurance Sector GIS uses

Insurance Sector is one of the prime sector which is indulged in the all sort of the sectors and individual as well. We all buy insurance policies for our own,for our vehicles,for our mobile phones or for any of our asset. We find insurance a very attractive thing which cover our losses. But have you ever thought that why the claim policies differs from one region to another.

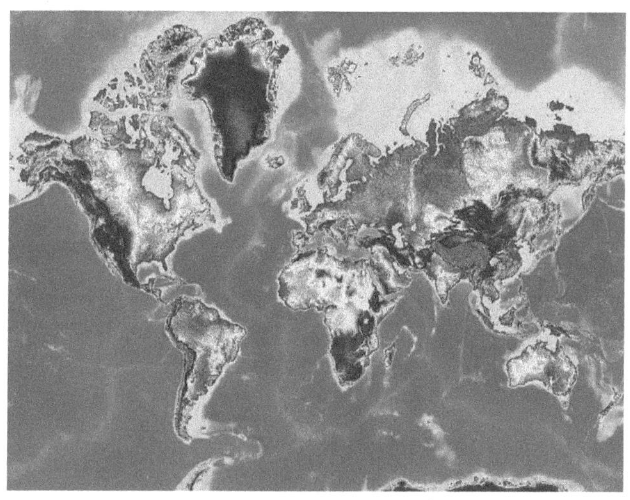

Insurance – GIS uses and application.

The reason for such differences occurs because of many factors out of which geographic conditions plays a major role. So, Insurance sector is highly driven by GIS. GIS helps this sector for following purpose:

- Disaster Prone Regions: With the help of GIS, Insurance providers assess the areas which are more disaster prone and what are the things which can be followed as to serve their clients.

- Risk assessment: Risk assessment is a major part of the insurance policies. People who provides insurance firstly assess the risk.They assess how much risk is included in any region and they make policies according to their assessment of risk.
- Claim management: Claim management is followed by the risk assessment and Risk management done with the help of GIS.
- Decision making tool: GIS works as a Decision making tool for the Insurance sector. They People use GIS for making decisions as to where to provide insurance and where to not. If providing it to any one then what should be the claim amount and so on.

Banking Sector GIS Application and Use

Banking sector helps in the proper flow of the money. Basically banking sector provides services for depositing money, providing loans, withdrawing money and what not. So basically the thing is to maintain the cash outflow and inflow in the bank and outer economy also.

Now days Banks are also into other activities and providing other services as well to generate the profits and making the customers look for more opportunities. So,all these activities becomes much easier with the help of the GIS.

GIS help Banks to maintain a good database of the customers. With the help of GIS,Banks can easily do the following process:

- Customer Database : GIS helps banks to maintain and visualize the data of customers as they will know from where we have the more customer,how many customers are from which region and belongs to which age group
- Market Analysis: Banks do analysis of the market trends and allow the people from different area and regions to bank with them.
- Focused Marketing: Focused marketing can be done very easily as banks can do the market analysis with the help of GIS. This Analysis can make Banks do marketing for the focused areas,regions,people.
- Better Services: GIS make Banks to serve best to their customer as to they can know the area where customer needs to be focused.

Logistics GIS Uses

Logistics, in a layman language we can call it as transportation. We all use transportation for moving our luggage,our vehicles and for many things. So, lets check how GIS can be useful for the Logistic service providers:

- Navigation: GIS can help the Logistic service providers in navigating from one place to another,showing them routes and the shortest and the best route to reach their destination.
- Cost Reduction: GIS helps in navigating and reducing cost also, GIS will make the route and will show the routes out of which you can choose any one whichever you find most suitable.

- Tracking: Logistics service providers and the customers can also track their orders with the help of GIS.

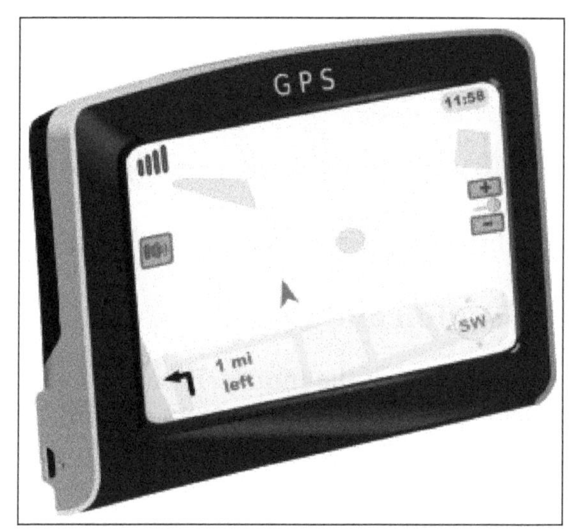

Tracking – GIS uses and application.

Disaster Management GIS Application and Uses

In every country there is a separate department for Disaster Management. This department is specially organized for looking over the Disaster and how to manage those disaster if any occurs.

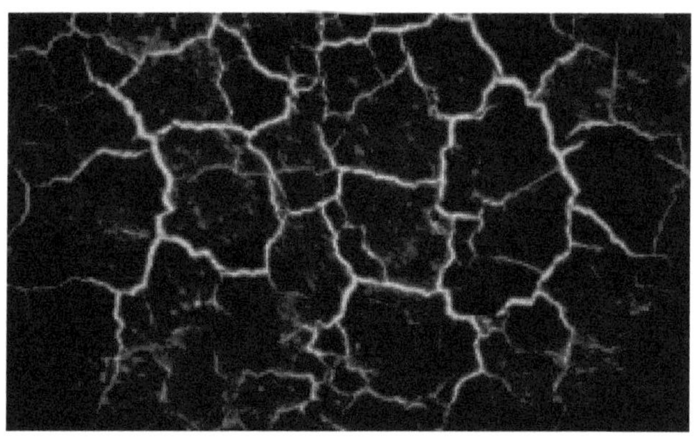

Disaster Management – GIS uses and application.

This department take very good care of all the disaster related things with the help of GIS, in the following way:

- Disaster Prone Area: The main and primary function of the Disaster management department is to identify the areas where disaster is more likely to occur. Disaster can be earthquake or flood or any miss happening.
- Planning of Rescue and evacuation: They identify the disaster prone areas with the help of GIS, that helps them to look for the rescue and evacuation. They plan how to rescue and evacuate the people from that disaster prone areas, what would be the procedure.
- No construction Site: This department makes sure that no construction site is established in these disaster prone areas.

- Rehabiliation and post disaster Management: Another main role performed by this department is the rehabiliation of the place and the people which has been effected because of the disaster. They make sure the situation comes back to normal as soon as possible.

Education Sector GIS Uses

Education is the basic necessity of the people now days, Education allow us to grow,explore, and it helps us to make a difference in the society. GIS is also giving something to education sector also.

- Degree: GIS is now allowing students to earn a degree in GIS. We know the course as Geo Spatial. Many universities are providing Geo spatial courses to its students.
- Difference in education system: Students are now looking for the different sectors to make their career in. Students are not looking for the same regular courses rather they are looking for something which would be different than the trending courses as well as should have good opportunities to grow and explore more. GIS is fulfilling this requirement with utmost efficiency which is definitely making a difference in our education system.

Travels and Tourism GIS uses and Application

Travels and tourism is one of the growing business in today's era. Travels and tourism sector is highly using GIS to show visualized places to their customers. We all approach to travels and tourism companies to provide us a package tour whenever we go to spend our vacations to a new place. Lets have a look on how Travels and Tourism business holders uses GIS:

Travel – GIS uses and application.

- Visualization of tourist spot: GIS helps the customers and business as well to visualize all the tourist spot which helps business to attract customers by actually giving them visualized view of the spots, which they would be visiting.
- Tourist location: GIS helps them to visualize location where they can visit.
- Route planning: GIS helps in the route planning also.
- Accommodation: GIS will help you to get the best accommodation facility.

- Cultural Events and Special attraction: For looking up to Cultural events and special attraction of any particular area GIS will be very Helpful.

BCP (Business Continuation Plan) GIS Application and Uses

BCP or business continuation plan is one of the plan which businesses are implementing in their business. Business continuation plan is one of the way to keep our business working unaffected in adverse conditions. Adverse conditions could be any natural disaster or a human made like terrorist attack. This plan helps in continuation of business working. So to plan BCP business take help of GIS for:

- Damage Assessment and Repair during Emergencies: GIS is helping to plan BCP by the assessment of the damage that can happen and how to cover up those damages in emergencies.
- Evacuation: Evacuation for any damage has to be a planned in prior. With the help of GIS we can make the route for evacuating people out of the affected area.
- Office Relocation: At the time of adversities, there might be chances that we need to relocate our office, until everything gets back to normal. With the help of GIS we can get the appropriate location for relocating the office.
- SCM: SCM or supply chain management is one of the key factor for any business, specially for product providers. They need to supply their products further to its end users. GIS is required to Plan SCM in adversities.
- Threat Assessment: GIS helps business to assess the threat level before establishing any business in any area, and also for the BCP they assess the level of threat.
- Weather Mapping: GIS helps in mapping the weather for any location. GIS can help you the best to know weather conditions,for your office location.

Infrastructure Development GIS Application

GIS is continuously contributing in the development of the society. So now GIS is also contributing in the Development of Infrastructure.

Real Estate – GIS uses and application.

- Transportation: GIS is helping in the development of transportation facilities as we all know that our success is highly depends on infrastructure development. So, in lieu to it every sector has to be developed.
- Communication: Communication is the way of conveying our thoughts to other person. So GIS is providing, the Communication service providers to visualize the network availability in the area.
- Facilities and amenities: GIS helps to develop infrastructure of the country or society by providing the data of facilities and amenities available in which area, which are the areas lacking behind in terms of facilities and amenities, and also to get the data of the facilities and amenities which are provided.
- Real Estate.

Crime Mapping GIS uses and Application

Today Crime is increasing day by day, people are being unsympathetic to others. Crime has taken a big toll irrespective of the place, be it a small village or a metro city. People are committing crime to satisfy their ego, to get their work done with a fear and so on. It gets very much difficult to investigate about the crime, people involved, place where it has been done and what not. So we have a solution for all these problems as the Investigating team use GIS for solving any case.

- Helps in Law Enforcement: GIS can help in enforcing law to the regions according to the crime committed in the area.
- Display Spatial Patterns of events: GIS helps in visualizing the type or pattern of crime committed in different areas.
- Investigating Serial Offense: GIS helps in investigating serial offense as we can reach to the culprit by tracking his location.

Advertising Industry GIS Application

Advertising – GIS uses and application.

In 21st century, marketing and advertising plays a major role in making a business famous and approachable to the people. But we need to understand the areas where we have to advertise or in other words we have to prioritize the users. We should know who are the people going to use your product. GIS can help in this task very easily. Advertising companies use GIS for following :

- Market Segmentation: Market segmentation is a process of identifying and dividing the market according to the usage of the product. We can divide the market by age,gender, regions as to identify the people for whom the product is useful and to know the usefulness of the product on map.
- Target Market: Once the market segmentation is done, we need to target the market. GIS can help us to Visualize the customers from the segmented market.

3D Campus Mapping GISUses and Application

3D campus mapping is the another use of GIS. GIS helps in making the 3D map for campus. These maps are easier to use and way more convenient than other Maps. 3D maps are useful as they provide better understanding of the campus and the building. It makes so easy to find out the locations with the most efficiency. For all these we need GIS to make the 3D map for a campus.

GIS AND ARCHAEOLOGY

Geographic Information Systems has been an important tool in archaeology since the early 1990s. Indeed, archaeologists were early adopters, users, and developers of GIS and GIScience, Geographic Information Science. The combination of GIS and archaeology has been considered a perfect match, since archaeology often involves the study of the spatial dimension of human behavior over time, and all archaeology carries a spatial component.

Since archaeology looks at the unfolding of historical events through geography, time and culture, the results of archaeological studies are rich in spatial information. GIS is adept at processing these large volumes of data, especially that which is geographically referenced. It is a cost effective, accurate and fast tool. The tools made available through GIS help in data collection, its storage and retrieval, its manipulation for customized circumstances and, finally, the display of the data so that it is visually comprehensible by the user. The most important aspect of GIS in archaeology lies, however, not in its use as a pure map-making tool, but in its capability to merge and analyse different types of data in order to create new information. The use of GIS in archaeology has changed not only the way archaeologists acquire and visualise data, but also the way in which archaeologists think about space itself. GIS has therefore become more of a science than an objective tool.

GIS in Survey

Survey and documentation are important to preservation and archaeology, and GIS makes this research and fieldwork efficient and precise. Research done using GIS capabilities is used as a decision making tool to prevent loss of relevant information that could impact archaeological sites and studies. It is a significant tool that contributes to regional planning and for cultural resource management to protect resources that are valuable through the acquisition and maintenance of data about historical sites.

In archaeology, GIS increases the ability to map and record data when it is used directly at the excavation site. This allows for immediate access to the data collected for analysis and visualization as an isolated study or it can be incorporated with other relevant data sources to help understand the site and its findings better.

The ability of GIS to model and predict likely archaeological sites is used by companies that are involved with utilizing vast tracts of land resources like the Department of Transportation. Section 106 of the National Preservation Act specifically requires historical sites as well as others to be assessed for impact through federally funded projects. Using GIS to assess archaeological sites that may exist or be of importance can be identified through predictive modeling. These studies and results are then used by the management to make relevant decisions and plan for future development. GIS makes this process less time consuming and more precise.

There are different processes and GIS functionalities that are used in archaeological research. Intrasite spatial analysis or distributional analysis of the information on the site helps in understanding the formation, process of change and in documentation of the site. This leads to research, analysis and conclusions. The old methods utilized for this provide limited exposure to the site and provide only a small picture of patterns over broad spaces. Predictive modeling is used through data acquisition like that of hydrography and hypsography to develop models along with archaeological data for better analysis. Point data in GIS is used to focus on point locations and to analyze trends in data sets or to interpolate scattered points. Density mapping is done for the analysis of location trends and interpolation is done to aid surface findings through the creation of surfaces through point data and is used to find occupied levels in a site. Aerial data is more commonly used. It focuses on the landscape and the region and helps interpret archaeological sites in their context and settings. Aerial data is analyzed through predictive modeling which is used to predict location of sites and material in a region. It is based on the available knowledge, method of prediction and on the actual results. This is used primarily in cultural resource management.

GIS in Analysis

GIS are able to store, manipulate and combine multiple data sets, making complex analyses of the landscape possible. Catchment analysis is the analysis of catchment areas, the region surrounding the site accessible with a given expenditure of time or effort. Viewshed analysis is the study of what regions surrounding the site are visible from that site. This has been used to interpret the relationship of sites to their social landscape. Simulation is a simplified representation of reality, attempting to model phenomena by identifying key variables and their interactions. This is used to think through problem formulation, as a means of testing hypothetical predictions, and also as a means to generate data.

In recent years, it has become clear that archaeologists will only be able to harvest the full potential of GIS or any other spatial technology if they become aware of the specific pitfalls and potentials inherent in the archaeological data and research process. Archaeoinformation science attempts to uncover and explore spatial and temporal patterns and properties in archaeology. Research towards a uniquely archaeological approach to information processing produces quantitative methods and computer software specifically geared towards archaeological problem solving and understanding.

GIS AND PUBLIC HEALTH

(GIS) combine computer-mapping capabilities with additional database management and data analysis tools. Commercial GIS systems are very powerful and have touched many applications and industries, including environmental science, urban planning, agricultural applications, and others.

Public health is another focus area that has made increasing use of GIS techniques. A strict definition of public health is difficult to pin down, as it is used in different ways by different groups. In general, public health differs from personal health in that it is (1) focused on the health of populations rather than of individuals, (2) focused more on prevention than on treatment, and (3) operates in a mainly governmental (rather than private) context. These efforts fall naturally within the domain of problems requiring use of spatial analysis as part of the solution, and GIS and other spatial analysis tools are therefore recognized as providing potentially transformational capabilities for public health efforts.

Public health informatics (PHI) is an emerging specialty which focuses on the application of information science and technology to public health practice and research. As part of that effort, a GIS – or more generally a spatial decision support system (SDSS) – offers improved geographic visualization techniques, leading to faster, better, and more robust understanding and decision-making capabilities in the public health arena.

For example, GIS displays have been used to show a clear relationship between clusters of emergent Hepatitis C cases and those of known intravenous drug users in Connecticut. Causality is difficult to prove conclusively – collocation does not establish causation – but confirmation of previously established causal relationships (like intravenous drug use and Hepatitis C) can strengthen acceptance of those relationships, as well as help to demonstrate the utility and reliability of GIS-related solution techniques. Conversely, showing the coincidence of potential causal factors with the ultimate effect can help suggest a potential causal relationship, thereby driving further investigation and analysis.

Alternately, GIS techniques have been used to show a lack of correlation between causes and effects or between different effects. For example, the distributions of both birth defects and infant mortality in Iowa were studied, and the researchers found no relationship in those data. This led to the conclusion that birth defects and infant mortality are likely unrelated, and are likely due to different causes and risk factors.

GIS can support public health in different ways as well. First and foremost, GIS displays can help inform proper understanding and drive better decisions. For example, elimination of health disparities is one of two primary goals of Healthy People 2010, one of the preeminent public health programs in existence today in the US. GIS can play a significant role in that effort, helping public health practitioners identify areas of disparities or inequities, and ideally helping them identify and develop solutions to address those shortcomings. GIS can also help researchers integrate disparate data from a wide variety of sources, and can even be used to enforce quality control measures on those data. Much public health data is still manually generated, and is therefore subject to human-generated mistakes and miscoding. For example, geographic analysis of health care data from North Carolina showed that just over 40% of the records contained errors of some sort in the geographic information (city, county, or zip code), errors that would have gone undetected without the visual displays provided by GIS. Correction of these errors led not only to more correct GIS displays, but also improved ALL analyses using those data.

Issues with GIS for Public Health

There are also concerns or issues with use of GIS tools for public health efforts. Chief among those is a concern for privacy and confidentiality of individuals. Public health is concerned about the health of the population as a whole, but must use data on the health of individuals to make many of those assessments, and protecting the privacy and confidentiality of those individuals is of paramount importance. Use of GIS displays and related databases raises the potential of compromising those privacy standards, so some precautions are necessary to avoid pinpointing individuals based on spatial data. For example, data may need to be aggregated to cover larger areas such as a zip code or county, helping to mask individual identities. Maps can also be constructed at smaller scales so that less detail is revealed. Alternately, key identifying features (such as the road and street network) can be left off the maps to mask exact location, or it may even be advisable to intentionally offset the location markers by some random amount if deemed necessary.

It is well established in the literature that statistical inference based on aggregated data can lead researchers to erroneous conclusions, suggesting relationships that in fact do not exist or obscuring relationships that do in fact exist. This issue is known as the modifiable areal unit problem. For example, New York public health officials worried that cancer clusters and causes would be misidentified after they were forced to post maps showing cancer cases by ZIP code on the internet. Their assertion was that ZIP codes were designed for a purpose unrelated to public health issues, and so use of these arbitrary boundaries might lead to inappropriate groupings and then to incorrect conclusions.

GIS AND AQUATIC SCIENCE

(GIS) has become an integral part of aquatic science and limnology. Water by its very nature is dynamic. Features associated with water are thus ever-changing. To be able to keep up with these changes, technological advancements have given scientists methods to enhance all aspects of scientific investigation, from satellite tracking of wildlife to computer mapping of habitats. Agencies like the US Geological Survey, US Fish and Wildlife Service as well as other federal and state agencies are utilizing GIS to aid in their conservation efforts.

GIS is being used in multiple fields of aquatic science from limnology, hydrology, aquatic botany, stream ecology, oceanography and marine biology. Applications include using satellite imagery to identify, monitor and mitigate habitat loss. Imagery can also show the condition of inaccessible areas. Scientists can track movements and develop a strategy to locate locations of concern. GIS can be used to track invasive species, endangered species, and population changes.

One of the advantages of the system is the availability for the information to be shared and updated at any time through the use of web-based data collection.

GIS and Fish

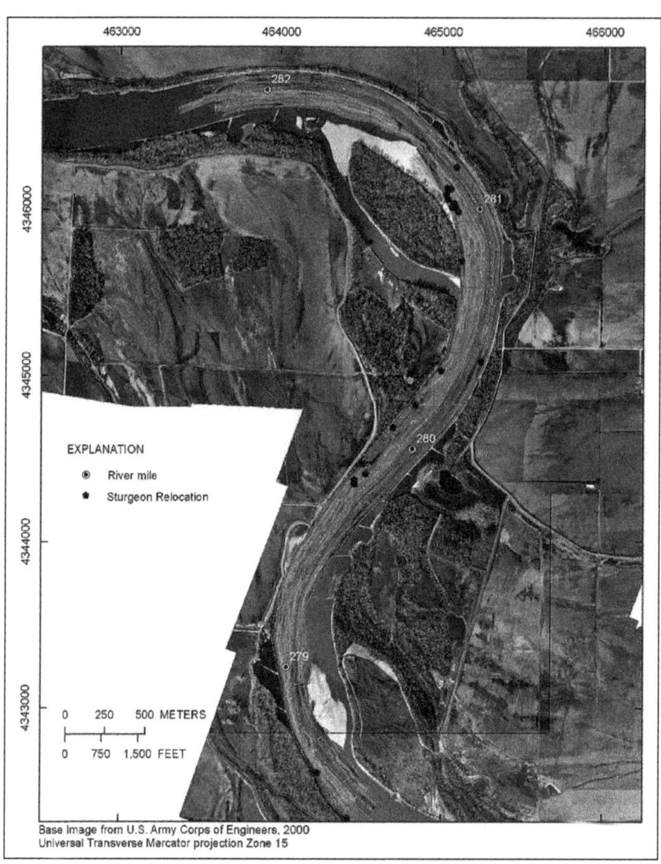

USGS sidescan radar image over base image from Army Corps of Engineers, indicating sturgeon location and river mile.

In the past, GIS was not a practical source of analysis due to the difficulty in obtaining spatial data on habitats or organisms in underwater environments. With the advancement of radio telemetry, hydroacoustic telemetry and side-scan sonar biologists have been able to track fish species and create databases that can be incorporated into a GIS program to create a geographical representation. Using radio and hydroacoustic telemetry, biologists are able to locate fish and acquire relatable data for those sites, this data may include substrate samples, temperature, and conductivity. Side-scan sonar allows biologists to map out a river bottom to gain a representation of possible habitats that are used. These two sets of data can be overlaid to delineate the distribution of fish and their habitats for fish. This method has been used in the study of the pallid sturgeon.

Over a period of time large amounts of data are collected and can be used to track patterns of migration, spawning locations and preferred habitat. Before, this data would be mapped and overlaid manually. Now this data can be entered into a GIS program and be layered, organized and analyzed in a way that was not possible to do in the past. Layering within a GIS program allows for the scientist to look at multiple species at once to find possible watersheds that are shared by these species, or to specifically choose one species for further examination. The US Geological Survey (USGS) in, cooperation with other agencies, were able to use GIS in helping map out habitat areas and movement patterns of pallid sturgeon. At the Columbia Environmental Research Center their effort relies on a customized ArcPad and ArcGIS, both ESRI (Environmental Systems Research

Institute) applications, to record sturgeon movements to streamline data collection. A relational database was developed to manage tabular data for each individual sturgeon, including initial capture and reproductive physiology. Movement maps can be created for individual sturgeon. These maps help track the movements of each sturgeon through space and time. This allowed these researchers to prioritize and schedule field personnel efforts to track, map, and recapture sturgeon.

GIS and Macrophytes

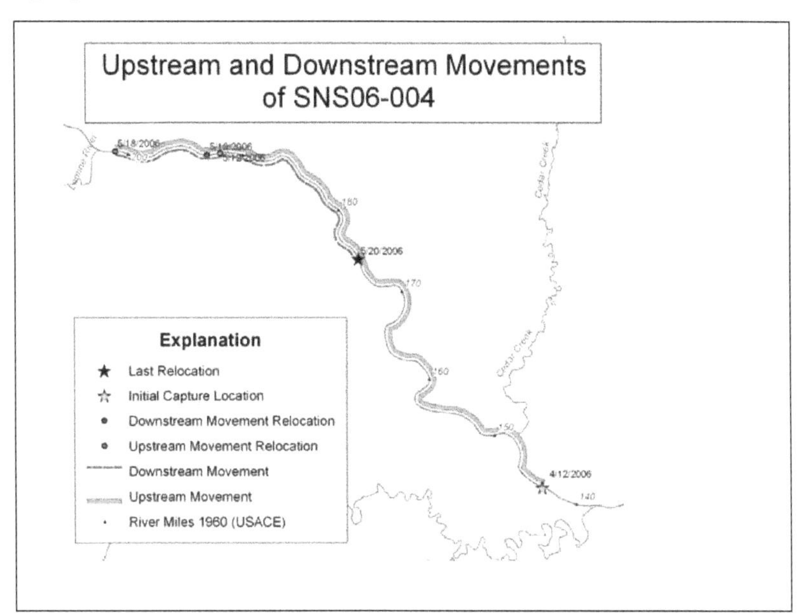

Map created from GIS database depicting the movements of individual sturgeon.

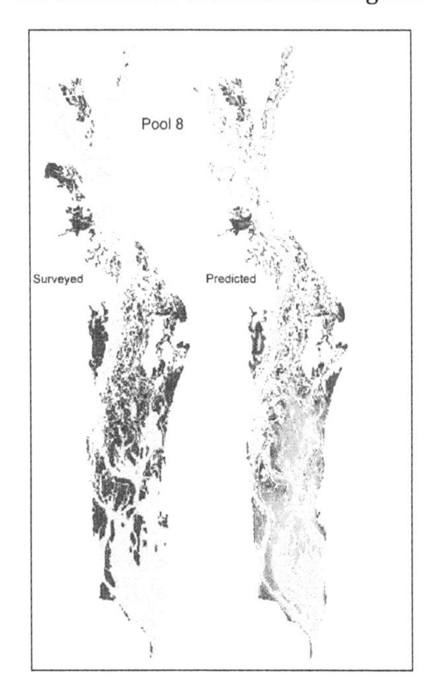

Surveyed (left) and predicted (right) distributions of submersed aquatic vegetation distribution Upper Mississippi River in 1989. The survey data were from the land cover/land use geographic information created by the U.S. Geological Survey Upper Midwest Environmental Sciences Center on the basis of interpretation of aerial photography of 1989.

Macrophytes are an important part of healthy ecosystems. They provide habitat, refuge, and food for fish, wildlife, and other organisms. Though natural occurring species are of great interest so are the invasive species that occur alongside these in our environment. GIS is being used by agencies and their respective resource managers as a tool to model these important macrophyte species. Through the use of GIS resource managers can assess the distributions of this important aspect of aquatic environments through a spatial and temporal scale. The ability to track vegetation change through time and space to make predictions about vegetation change are some of the many possibilities of GIS. Accurate maps of the aquatic plant distribution within an aquatic ecosystem are an essential part resource management.

It is possible to predict the possible occurrences of aquatic vegetation. For example, the USGS has created a model for the American wild celery (Vallisneria americana) by developing a statistical model that calculates the probability of submersed aquatic vegetation. They established a web link to an Environmental Systems Research Institute (ESRI) ArcGIS Server website *Submersed Aquatic Vegetation Model to make their model predictions available online. These predictions for distribution of submerged aquatic vegetation can potentially have an effect on foraging birds by creating avoidance zones by humans. If it is known where these areas are, birds can be left alone to feed undisturbed. When there are years where the aquatic vegetation is predicted to be limited in these important wildlife habitats, managers can be alerted.

Invasive species have become a great conservation concern for resource managers. GIS allows managers to map out plant locations and abundances. These maps can then be used to determine the threat of these invasive plants and help the managers decide on management strategies. Surveys of these species can be conducted and then downloaded into a GIS system. Coupled with this, native species can be included to determine how these communities respond with each other. By using known data of preexisting invasive species GIS models could predict future outbreaks by comparing biological factors. The Connecticut Agricultural Experiment Station Invasive Aquatic Species Program (CAES IAPP) is using GIS to evaluate risk factors. GIS allows managers to georeference plant locations and abundance. This allows for managers to display invasive communities alongside native species for study and management.

GIS AND ENVIRONMENT

(GIS) is a commonly used tool for environmental management, modelling and planning. As simply defined by Michael Goodchild, GIS is as 'a computer system for handling geographic information in a digital form'. In recent years it has played an integral role in participatory, collaborative and open data philosophies. Social and technological evolutions have elevated 'digital' and 'environmental' agendas to the forefront of public policy, the global media and the private sector.

Government departments routinely use digital spatial platforms to plan and model proposed changes to road networks, building design, greenbelt land, utility provision, crime prevention, energy production, waste management and security. Non-profit organizations also incorporate geospatial and web-mapping approaches into political campaigns to lobby governments, to protest against socially or environmentally harmful companies, and to generate public support. Private business, whether in land management, resource extraction, retail, manufacturing or social media for example, also incorporate GIS into overall profit-making strategies.

Citizen Science and GIS

Citizen science is part of the wider emphasis upon public involvement in expert fields across Western democracies. The term is 'often used to describe communities or networks of citizens who act as observers in some domain of science'. Although more narrowly used to describe the shift to specifically user-generated forms of knowledge creation, it has been routinely invoked in both the public participatory GIS and environmental governance literature at large.

The Secretary of the US Navy, Ray Mabus is briefed on the Deepwater Horizon oil spill response. A web-based GIS is visible in the background. The NOAA-developed Environmental Response Management Application (ERMA) was designed to assist resource managers post-spill.

Mapping for Change

Another example of citizen science and GIS in action is taken from inside the academy. University College London (UCL) and London 21 sustainability network's Mapping for Change initiative has encouraged voluntary groups, local authorities and development agencies to build map-based projects to support political, social and environmental aims. They even provide a noise mapping toolkit on the Mapping for Change website itself, designed to help local communities gather evidence of intrusive and unwanted environmental nuisances and hazards. The Royal Docks community in London has used such a toolkit to help present their concerns to the Greater London Authority Environment Committee over plans to expand London City Airport. Armed with sound meters, survey sheets and access to an online mapping platform, residents were able to monitor noise levels; from overhead planes and passing motor vehicles, to birdsong and ambient river sounds.

A plane lands at London City Airport. Communities in the area have used web-based noise maps to demonstrate their objections to proposed flight expansion.

Their data was then visualized in various formats to help advance their argument. Royal Docks' residents are continually plagued by planes taking-off and landing at London City Airport, and plans to expand the number of flights a year by 50% (up to 120,000) were opposed by local communities on the basis that it would decrease their quality of life.

GIS and citizen science go hand-in-hand. Web-based mapping platforms serve as useful tools for national conservation societies, local community groups and planning departments to compile tangible data on environmental issues. Voluntary, grassroots approaches can help compile lay knowledges that feed back into more formal political frameworks.

Environmental Justice

At a local level, GIS has been frequently used to engage stakeholders in the planning of environmentally 'bad' sites. Nuclear power stations, wind farms, landfill sites, and other energy facilities are often subject to NIMBY opposition for aesthetic, health and social reasons. This is despite of their capacity to produce 'good' economic factors or employment opportunities. GIS has thus found itself deployed alongside broader cost-benefit analysis (CBA), and multi-criteria decision analysis (MCDA) approaches to socio-political conflict. Environmental Justice (EJ) activists believe these decisions act to further embed racial and class divides. GIS provides an important angle to the EJ movement.

Elements of Justice

An aerial photograph of New Orleans after Hurricane Katrina in 2005. The High magnitude flood event disproportionately affected the city's poor, black neighbourhoods.

Broadly, the EJ movement is a loose connection of social groups, stakeholders and activists who have sought to contest socio-political injustices. Commonly, this has been through a single motive; the equal distribution of environmental goods and bads. As Schlosberg contends, 'the issue of distribution is always present and always key' to the guiding EJ ethos. Yet, other demands are frequently put forward. Following Schlosberg, there are two further demands that constitute the EJ movement than a mere 'redistribution of environmental ills and benefits. The first is the 'recognition of the diversity of participants and experiences in affected communities'. Thus, EJ demands that people affected by environmental injustices are appropriately noticed by others. A lack of recognition in local community discourses, 'demonstrated by various forms of insults, degradation, and devaluation', marginalize those already least able to contest political decisions. The second is the notion of *participatory justice*. According to Schlosberg; 'if you are not recognised, you do not participate.' Thus, recognitional justice leads directly to participatory justice. In EJ terms, participation is about involving those outside the typical political/institutional order. Democratic and participatory decision-making procedures are both an element of, and a condition for, social justice. Simultaneously, institutionalised exclusion, social cultures of misrecognition, and current distributional patterns can be challenged.

New Urban Landscapes

GIS has also had a role in formulating new urban landscapes. Planned cities - designed entirely from scratch - routinely use digital technologies to visualize and demonstrate urban layouts, building structures and transport arrangements. Although CAD/CAM technologies are often used to assist in the visualization, construction, and delivery of certain engineering features, GIS helps to realize distinctly spatial components of the city. Environmental narratives of a 'carbon-free' and sustainable future favour those in the GIS industry. 'The challenge of the 21st century' as ESRI would have it, 'is to arrest the progress of climate change'. Geospatial software has played its part in developing this narrative.

Masdar City

Masdar City is a 'sustainable, zero-carbon, zero-waste' project currently under construction in the United Arab Emirates (UAE). Situated in the Abu Dhabi emirate, Masdar is described as 'an emerging global hub for renewable energy and clean technologies'. The Abu Dhabi Future Energy Company have funded and overseen the $18bn dollar project. No cars are allowed on its streets, energy is produced in part by renewable sources, building materials are 'sustainable' and water-use is controlled. GIS is being employed to plan, facilitate and test a plethora of environmental phenomena and technological processes.

A dedicated GIS team is responsible for 'managing the overall spatial information needs' of the project, starting with the drawing of a common base map with which to support the city's infrastructure. Without a spatial plan of Masdar's operative mechanisms, the city will fail to deliver its grand ambition. In particular, GIS is being used to visualize, analyse and model *land-use* in the city. Masdar – unlike any other city – has to incorporate a wealth of energy-related facilities within its perimeter. As EJ activists are all too aware, the siting of such facilities can be a key area of conflict. Masdar's water treatment and sewage plants, material recycling centre, solar power plant, geothermal test site, solar panel test field and concrete batching plant all need to be situated inside the city's boundaries. As CH2M HILL's Site Control and GIS Manager for the Masdar project confirms; 'never have so many environmental facilities come together in one place'. GIS is the central tool with which to imagine – in a digital environment – different siting scenarios. In this case, GIS is seen to operate as a decision-making tool; informing the practitioners who work on the Masdar project.

GIS is also being used to model some of Masdar's key infrastructural features directly. Its involvement in simulating the citywide Personal Rapid Transport System (PRTS) is one such example. As common road vehicles are banned from the city, the 'driverless' transit system will transport people and freight across the 7km2 area. GIS is capable of modelling the system route, due to comprise 85 passenger stations and approximately 1,700 automated vehicles. By drawing spatial buffer zones around potential PRTS stops, passenger-distance maps can visualize residential areas that fall outside of ideal service requirements. GIS is an instrumental tool in visualizing such problems. A smooth, functioning PRTS is a central infrastructural aspect of Masdar's grand vision, and engineering companies who specialize in GIS technologies have helped in realizing this digitally-orchestrated dream. Yet journalists in particular have been sceptical. As Bryan Walsh has argued; will Masdar City ever really develop the authenticity of a real city?'. Or as Jonathan Glancey contends, will its 'ultra-modern aspects prove to be a mirage'?

Post-political Agendas

'Post-politics' is a neologism for the consensual, participatory and techno-managerial approach to modern governance. Originally coined by Slavoj Zizek, the post-political critique argues that life in the Western world is routinely characterized by the de-politicizing effects of a 'consensual police order'. A number of different techno-managerial 'fixes' have been sought by neoliberal governments in order to solve expressly environmental problems, rather than due political processes. As Swyngedouw has argued, such forces have 'replaced debate, disagreement and dissensus with a series of technologies of governing that fuse around consensus, agreement, accountancy merits and technocratic environmental management'. Thus if the rise of the post-political order is due to the increasing reliance upon 'technocratic environmental management', as Swyngedouw has argued, then GIS – as a tool for neoliberal environmental governance – is implicit in such an order.

The 2009 Copenhagen Summit failed to produce a legally binding treaty on national CO_2 levels, despite (or perhaps because of) a near-global delegation. An example of Jacques Ranciere's partition of the sensible in action. IPCC data, based on GIS and environmental modelling processes, provide the basis for these proposed legal frameworks.

The Police, The Political and Politics

Firstly, it serves to elucidate upon claims to a 'post-political impasse'. This is best understood through what Ranciere calls *the police*, *the political* and *politics*. It is within these three terms that Ranciere carves out what he calls the true meaning of political action, and of what it is to exercise political right. As Panagia neatly summarizes; 'politics is the practice of asserting one's position that ruptures the logic of arche'. Politics is about rediscovering the art of debate, conflict and struggle, and not merely about re-organizing the administrative framework of existing political structures (namely, the state apparatus). *The political* – in Ranciere's words – is for 'the one who is 'unaccounted-for', the one who has no speech to be heard'. Democracy does not work towards an 'idealized-normative condition' of equal rights, but is built upon the very ontological notion of such. *Politics* brings the political into the foreground; rendering that which was previously noise, legitimate speech. Much of what is thought to be politics in the contemporary world is actually subsumed within *the police*. Policy is not politics is this sense, then. This is what Ranciere calls *the partition of the sensible*, or the established order of things.

The post-politics of contemporary governance brings all that it can into this order. Those who were previously cast outside the police structure are now "responsible' partners' in a stakeholder-based arrangement. All views that were previously antagonistic and conflictual are now brought together in a more homely, consensual arena. No room is made for 'irrational' demands. As such, Ranciere says that, 'consensus is the reduction of politics to the police'. Nowhere has this been more visible than within the apocalyptic climate narratives told *ad infinitum* by environmental practitioners, policy-makers and non-governmental organizations. Consensual politics has found its home in the environmental arena.

Post-political GIS

In short, GIS works as a tool for mediating and diffusing socio-environmental conflicts. It does so by working within Ranciere's notion of the partition of the sensible. Whilst it may allow previously unheard voices to gain a voice (in environmental justice campaigns, foremostly), it still does not – as a tool of neoliberal governance – make room for those who are deemed 'outside', unruly or conflictual to have a voice. For example, Elwood laments the notionally 'participatory' flag-waving carried out by those involved in urban GIS-based projects. As she says, 'the skills and financial and temporal costs of using GIS effectively bar many individuals, social groups and organizations from participation in research and decision-making where it is used', denying those without the means to participate, from participation. GIS does not necessarily facilitate involvement for all.

Moreover, GIS is limited in its ontological scope, reducing all things spatial to a *calculable order*. As Leszczynski contends, GIS operates a 'discourse populated by discrete objects of knowledge'; differentiating 'between the binary of truth and error'. GIS is thus a central *polic-ing* tool for contemporary socio-environmental governance. It works to order space into discrete and ordered formats.

GOOGLE EARTH

Google Earth is a computer program that renders a 3D representation of Earth based primarily on satellite imagery. The program maps the Earth by superimposing satellite images, aerial photography, and GIS data onto a 3D globe, allowing users to see cities and landscapes from various angles. Users can explore the globe by entering addresses and coordinates, or by using a keyboard or mouse. The program can also be downloaded on a smartphone or tablet, using a touch screen or stylus to navigate. Users may use the program to add their own data using Keyhole Markup

Language and upload them through various sources, such as forums or blogs. Google Earth is able to show various kinds of images overlaid on the surface of the earth and is also a Web Map Service client.

In addition to Earth navigation, Google Earth provides a series of other tools through the desktop application. Additional globes for the Moon and Mars are available, as well as a tool for viewing the night sky. A flight simulator game is also included. Other features allow users to view photos from various places uploaded to Panoramio, information provided by Wikipedia on some locations, and Street View imagery. The web-based version of Google Earth also includes Voyager, a feature that periodically adds in-program tours, often presented by scientists and documentarians.

Google Earth has been viewed by some as a threat to privacy and national security, leading to the program being banned in multiple countries. Some countries have requested that certain areas be obscured in Google's satellite images, usually areas containing military facilities.

Imagery

Google Earth's imagery is displayed on a digital globe, which displays the planet's surface using a single composited image from a far distance. After zooming in far enough, the imagery transitions into different imagery of the same area with finer detail, which varies in date and time from one area to the next. The imagery is retrieved from satellites or aircraft. Before the launch of NASA and the USGS's Landsat 8 satellite, Google relied partially on imagery from Landsat 7, which suffered from a hardware malfunction that left diagonal gaps in images. In 2013, Google used datamining to remedy the issue, providing what was described as a successor to the Blue Marble image of Earth, with a single large image of the entire planet. This was achieved by combining multiple sets of imagery taken from Landsat 7 to eliminate clouds and diagonal gaps, creating a single "mosaic" image. Google now uses Landsat 8 to provide imagery in a higher quality and with greater frequency. Imagery is hosted on Google's servers, which are contacted by the application when opened, requiring an Internet connection.

Imagery resolution ranges from 15 meters of resolution to 15 centimeters. For much of the Earth, Google Earth uses digital elevation model data collected by NASA's Shuttle Radar Topography Mission. This creates the impression of three-dimensional terrain, even where the imagery is only two-dimensional.

Every image created from Google Earth using satellite data provided by Google Earth is a copyrighted map. Any derivative from Google Earth is made from copyrighted data which, under United States Copyright Law, may not be used except under the licenses Google provides. Google allows non-commercial personal use of the images (e.g. on a personal website or blog) as long as copyrights and attributions are preserved. By contrast, images created with NASA's globe software World Wind use The Blue Marble, Landsat, or USGS imagery, each of which is in the public domain.

In version 5.0, Google introduced Historical Imagery, allowing users to view earlier imagery. Clicking the clock icon in the toolbar opens a time slider, which marks the time of available imagery from the past. This feature allows for observation of an area's changes over time. Utilizing the timelapse feature allows for the ability to view a zoomable video as far back as 32 years.

3D Imagery

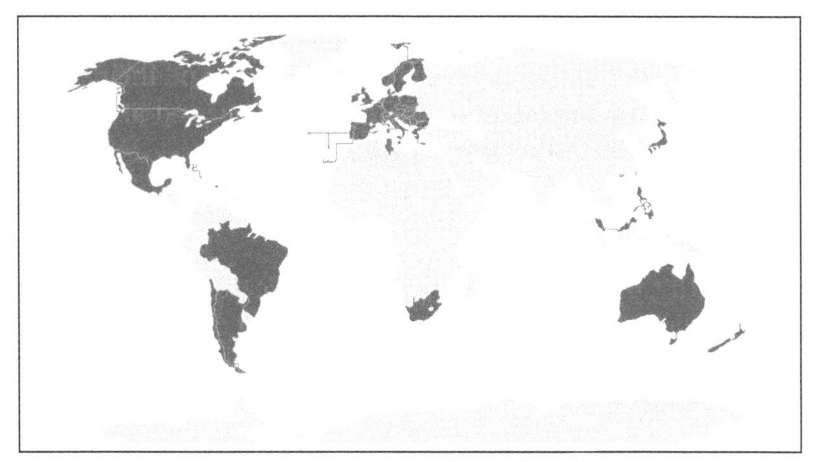

Countries with 3D coverage in Google Earth as of February 2019.

3D imagery in the iOS version of Google Earth, seen here at Wisconsin Dells, Wisconsin.

Google Earth shows 3D building models in some cities, including photorealistic 3D imagery. The first 3D buildings in Google Earth were created using 3D modeling applications such as SketchUp and, beginning in 2009, Building Maker, and were uploaded to Google Earth via the 3D Warehouse. In June 2012, Google announced that it would be replacing user-generated 3D buildings with an auto-generated 3D mesh. This would be phased in, starting with select larger cities, with the notable exception of cities such as London and Toronto which required more time to process detailed imagery of their vast number of buildings. The reason given is to have greater uniformity in 3D buildings, and to compete with Nokia Here and Apple Maps, which were already using this technology. The coverage began that year in 21 cities in four countries. By early 2016, 3D imagery had been expanded to hundreds of cities in over 40 countries, including every U.S. state and encompassing every continent except Antarctica.

In 2009, in a collaboration between Google and the Museo del Prado in Madrid, the museum selected 14 of its paintings to be photographed and displayed at the resolution of 14,000 megapixels inside the 3D version of the Prado in Google Earth and Google Maps.

Street View

On April 15, 2008, with version 4.3, Google fully integrated Street View into Google Earth. Street View displays 360° panoramic street-level photos of select cities and their surroundings. The photos were taken by cameras mounted on automobiles, can be viewed at different scales and from many angles, and are navigable by arrow icons imposed on them.

Water and Ocean

Introduced in Google Earth 5.0 in 2009, the Google Ocean feature allows users to zoom below the surface of the ocean and view the 3D bathymetry. Supporting over 20 content layers, it contains information from leading scientists and oceanographers. On April 14, 2009, Google added bathymetric data for the Great Lakes.

In June 2011, Google increased the resolution of some deep ocean floor areas from 1-kilometer grids to 100 meters. The high-resolution features were developed by oceanographers at Columbia University's Lamont-Doherty Earth Observatory from scientific data collected on research cruises. The sharper focus is available for about 5 percent of the oceans. This can be seen in the Hudson off New York City, the Wini Seamount near Hawaii, and the Mendocino Ridge off the U.S Pacific coast.

Outer Space

A picture of Mars' landscape.

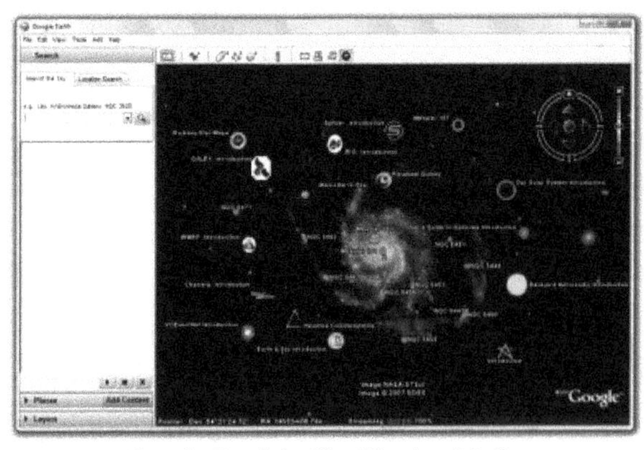

Google Earth in Sky Viewing Mode.

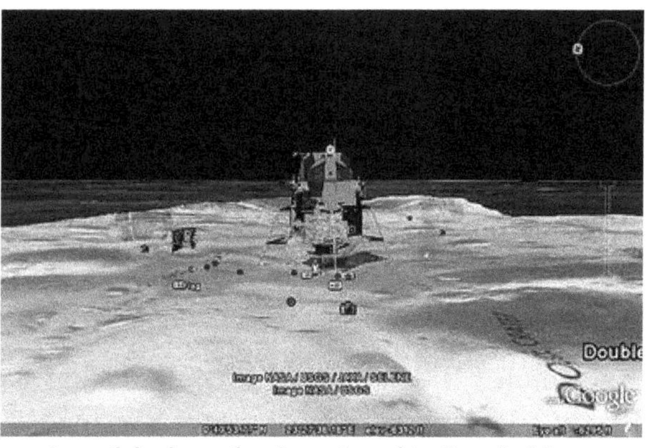

One of the lunar landers viewed in Google Moon.

Google has programs and features, including within Google Earth, allowing exploration of Mars, the Moon, the view of the sky from Earth and outer space, including the surfaces of various objects in the Solar System.

Google Sky

Google Sky is a feature that was introduced in Google Earth 4.2 on August 22, 2007, in a browser-based application on March 13, 2008, and to Android smartphones, with augmented reality features. Google Sky allows users to view stars and other celestial bodies. It was produced by Google through a partnership with the Space Telescope Science Institute (STScI) in Baltimore, the science operations center for the Hubble Space Telescope. Dr. Alberto Conti and his co-developer Dr. Carol Christian of STScI plan to add the public images from 2007, as well as color images of all of the archived data from Hubble's Advanced Camera for Surveys. Newly released Hubble pictures will be added to the Google Sky program as soon as they are issued.

New features such as multi-wavelength data, positions of major satellites and their orbits as well as educational resources will be provided to the Google Earth community and also through Christian and Conti's website for Sky. Also visible on Sky mode are constellations, stars, galaxies, and animations depicting the planets in their orbits. A real-time Google Sky mashup of recent astronomical transients, using the VOEvent protocol, is being provided by the VOEventNet collaboration. Other programs similar to Google Sky include Microsoft WorldWide Telescope and Stellarium.

Google Mars

Google Mars is an application within Google Earth that is a version of the program for imagery of the planet Mars. Google also operates a browser-based version, although the maps are of a much higher resolution within Google Earth, and include 3D terrain, as well as infrared imagery and elevation data. There are also some extremely-high-resolution images from the Mars Reconnaissance Orbiter's HiRISE camera that are of a similar resolution to those of the cities on Earth. Finally, there are many high-resolution panoramic images from various Mars landers, such as the Mars Exploration Rovers, *Spirit* and *Opportunity*, that can be viewed in a similar way to Google Street View.

Mars also has a small application found near the face on Mars. It is called Meliza, a robot character the user can speak with.

Google Moon

Originally a browser application, Google Moon is a feature that allows exploration of the Moon. Google brought the feature to Google Earth for the 40th anniversary of the Apollo 11 mission on July 20, 2009. It was announced and demonstrated to a group of invited guests by Google along with Buzz Aldrin at the Newseum in Washington, D.C. Google Moon includes several tours, including one for the Apollo missions, incorporating maps, videos, and Street View-style panoramas, all provided by NASA.

Other Features

Google Earth has numerous features which allow the user to learn about specific places. These are called "layers", and include different forms of media, including photo and video. Some layers include tours, which guide the user between specific places in a set order. Layers are created using the Keyhole Markup Language, or KML, which users can also use to create customized layers. Locations can be marked with placemarks and organized in folders; For example, a user can use placemarks to list interesting landmarks around the globe, then provide a description with photos and videos, which can be viewed by clicking on the placemarks while viewing the new layer in the application.

In December 2006, Google Earth added a new integration with Wikipedia and Panoramio. For the Wikipedia layer, entries are scraped for coordinates via the Coord templates. There is also a community-layer from the project Wikipedia-World. More coordinates are used, different types are in the display, and different languages are supported than the built-in Wikipedia layer. The Panoramio layer features pictures uploaded by Panoramio users, placed in Google Earth based on user-provided location data. In addition to flat images, Google Earth also includes a layer for user-submitted panoramic photos, navigable in a similar way to Street View.

Google Earth includes multiple features that allow the user to monitor current events. In 2007, Google began offering users the ability to monitor traffic data provided by Google Traffic in real time, based on information crowdsourced from the GPS-identified locations of cell phone users.

Flight Simulators

Downtown Toronto as seen from an F-16 Fighting Falcon during a simulated flight.

In Google Earth 4.2, a flight simulator was added to the application. It was originally a hidden feature when introduced in 2007, but starting with 4.3, it was given a labeled option in the menu. In addition to keyboard control, the simulator can be controlled with a mouse or joystick. The simulator also runs with animation, allowing objects such as planes to animate while on the simulator.

Another flight simulator, GeoFS, was created under the name GEFS-Online using the Google Earth Plug-in API to operate within a web browser. As of September 1, 2015, the program now uses the open-source program CesiumJS, due to the Google Earth Plug-in being discontinued.

Liquid Galaxy

Liquid Galaxy is a cluster of computers running Google Earth creating an immersive experience. On September 30, 2010, Google made the configuration and schematics for their rigs public, placing code and setup guides on the Liquid Galaxy wiki. Liquid Galaxy has also been used as a panoramic photo viewer using KRpano, as well as a Google Street View viewer using Peruse-a-Rue Peruse-a-Rue is a method for synchronizing multiple Maps API clients.

GEOGRAPHIC INFORMATION SYSTEMS IN GEOSPATIAL INTELLIGENCE

Geographic information systems (GIS) play a constantly evolving role in geospatial intelligence (GEOINT) and United States national security. These technologies allow a user to efficiently manage, analyze, and produce geospatial data, to combine GEOINT with other forms of intelligence collection, and to perform highly developed analysis and visual production of geospatial data. Therefore, GIS produces up-to-date and more reliable GEOINT to reduce uncertainty for a decisionmaker. Since GIS programs are Web-enabled, a user can constantly work with a decision maker to solve their GEOINT and national security related problems from anywhere in the world. There are many types of GIS software used in GEOINT and national security, such as Google Earth, ERDAS IMAGINE, GeoNetwork opensource, and Esri ArcGIS.

Geospatial Intelligence (GEOINT)

GEOINT, known previously as imagery intelligence (IMINT), is an intelligence collection discipline that applies to national security intelligence, law enforcement intelligence, and competitive intelligence. For example, an analyst can use GEOINT to identify the route of least resistance for a military force in a hostile country, to discover a pattern in the locations of reported burglaries in a neighborhood, or to generate a map and comparison of failing businesses that a company is likely to purchase. GEOINT is also the geospatial product of a process that is focused externally, designed to reduce the level of uncertainty for a decisionmaker, and that uses information derived from all sources. The National Geospatial-Intelligence Agency (NGA), who has overall responsibility for GEOINT in the U.S. Intelligence Community (IC), defines GEOINT as "information about any object—natural or man-made—that can be observed or referenced to the Earth, and has national security implications."

Some of the sources of collected imagery information for GEOINT are imagery satellites, cameras on airplanes, Unmanned Aerial Vehicles (UAV) and drones, handheld cameras, maps, or GPS coordinates. Recently the NGA and IC have increased the use of commercial satellite imagery for intelligence support, such as the use of the IKONOS, Landsat, or SPOT satellites. These sources produce digital imagery via electro-optical systems, radar, infrared, visible light, multispectral, or hyperspectral imageries.

The advantages of GEOINT are that imagery is easily consumable and understood by a decisionmaker, has low human life risk, displays the capabilities of a target and its geographical relationship to other objects, and that analysts can use imagery world-wide in a short time. On the other hand, the disadvantages of GEOINT are that imagery is only a snapshot of a moment in time, can be too compelling and lead to ill-informed decisions that ignore other intelligence, is static and vulnerable to deception and decoys, does not depict the intentions of a target, and is expensive and subject to environmental problems.

GIS use in GEOINT and National Security Intelligence

A majority of national security intelligence decisions involve geography and GEOINT. GIS allows the user to capture, manage, exploit, analyze, and visualize geographically referenced information, physical features, and other geospatial data. GIS is thus a critical infrastructure for the GEOINT and national security community in manipulating and interpreting spatial knowledge in an information system. GIS extracts real world geographic or other information into datasets, maps, metadata, data models, and workflow models within a geodatabase that is used to solve GEOINT-related problems. GIS provides a structure for map and data production that allows a user to add other data sources, such as satellite or UAV imagery, as new layers to a geodatabase. The geodatabase can be disseminated and operated across any network of associated users (i.e. from the GEOINT analyst to the warfighter) and engenders a common spatial capability for all defense and intelligence domains.

The map and chart production agency and imagery intelligence agency, the principal two agencies of GEOINT, use GIS to efficiently work together to solve decisionmaker's geospatial questions, to communicate effectively between their unique departments, and to provide constantly updated, accurate GEOINT to their national security and warfighter domains.

Another important aspect of GIS is its ability to fuse geospatial data with other forms of intelligence collection, such as signals intelligence (SIGINT), measurement and signature intelligence (MASINT), human intelligence (HUMINT), or open source intelligence (OSINT). A GIS user can incorporate and fuse all of these types of intelligence into applications that provide corroborated GEOINT throughout an organization's information system.

GIS enables efficient management of geospatial data, the fusion of geospatial data with other forms of intelligence collection, and advanced analysis and visual production of geospatial data. This produces faster, corroborated, and more reliable GEOINT that aims to reduce uncertainty for a decisionmaker.

Roles

- Data and map production.
- Data fusion, data discovery through metadata catalogs, and data dissemination through Web portals and browsers.
- Analysis and exploitation of collected imagery or intelligence.
- SIGINT, GEOINT, MASINT, and other sensor analysis.
- Fusion of multiple forms of intelligence collection.
- Collaborative planning and efficient workflow management between decisionmakers, analysts, consumers, and warfighters.
- Suitability and temporal analysis.
- Stewardship: Geospatial intelligence.

Related Esri Products

Distributed Geospatial Intelligence Network (DGInet)

The DGInet technology allows military and national security intelligence customers to access large multi-terabyte databases through a common Web-based interface. This gives the users the capability to quickly and easily identify, overlay, and fuse georeferenced data from various sources to create maps or support geospatial analysis. Esri designed the technology for inexperienced GIS users of national security intelligence and defense organizations in order to provide a Web-based enterprise solution for publishing, distributing, and exploiting GEOINT data among designated organizations. According to Esri, the DGInet technology "uses thin clients to search massive amounts of geospatial and intelligence data using low-bandwidth Web services for data discovery, dissemination, and horizontal fusion of data and products."

PLTS for ArcGIS Specialized Solutions

PLTS for ArcGIS Specialized Solutions is a group of software applications that extends ArcGIS to facilitate database driven cartographic production for geospatial and mapping agencies, nautical and aeronautical chart production, foundation mapping, and defense mapping requirements. The collection of software applications includes Esri Production Mapping, Esri Nautical Solution, Esri Aeronautical Solution, and Esri Defense Mapping programs that provide quality control, easier and consistent map production, database sharing, and efficient workflow management for each program's specific type of mapping or charting.

Geoprocessing

Geoprocessing is based on a framework of data transformation in GIS and is a collection of hundreds of GIS tools that manipulate geospatial or other data in GIS. A geoprocessing tool performs an operation (often the name of the tool, such as "Clip") on an existing GIS dataset and produces a new dataset as a result of the utilized tool. GIS users utilize these tools to create a workflow model that quickly and easily transforms raw data into the desired product.

In GEOINT, users employ geoprocessing in similar ways. They can make geoprocessing tools resemble analytic techniques to transform large amounts of data into actionable information. In national security intelligence and defense organizations, geoprocessing notifies users to events occurring in specific areas of interest and enables domain-specific analysis applications, such as radio frequency analysis, terrain analysis, and network analysis.

Tracking Analyst and Tracking Server

The ArcGIS Tracking Analyst extension enables the user to create time series visualizations to analyze time and location sensitive information. It creates a visible path from incorporated data that shows movement through space and time. The program allows the national security intelligence or defense user to track assets (such as vehicles or personnel), monitor sensors, visualize change over time, play back events, and analyze historical or real-time temporal data.

The Tracking Server program is an Esri enterprise technology that integrates real-time data with GIS to disseminate information quickly and easily to decisionmakers. This program enables the user to obtain data in any format and transmit it to the necessary consumer or ArcGIS Tracking Analyst user, to conduct filters or alerts on specific attributes of incoming data or global positions, and to log data into ArcGIS Server for efficient project management and information sharing.

When Tracking Server and ArcGIS Tracking Analyst are used together, a user can monitor changes in data as they occur in real-time. A national security intelligence or defense user can subscribe to real-time data over the Internet from GPS and custom data feeds to support GEOINT requirements, such as fleet management or target tracking.

ArcGIS Military Analyst

The ArcGIS Military Analyst extension incorporates display and analysis tools that allow the use and production of vector and raster products, line-of-sight analysis, hillshade analysis, terrain analysis, and Military Grid Reference System (MGRS) conversion. This program also provides a basis for command, control, and intelligence (C2I) systems. National security intelligence and defense organizations use ArcGIS Military Analyst extension to integrate geospatial data with other defense data, analyze digital terrains, and prepare for battle. This program also enables such users to manage and analyze geospatial data and relationships between mission planning, logistics, and C2I.

Military Overlay Editor (MOLE)

MOLE is a set of command components that enables national security intelligence and defense users to easily create, display, and edit U.S. Department of Defense MIL-STD-2525B and the North Atlantic Treaty Organization APP-6A military symbology in a map. This allows for easier and faster identification, understanding, and movement of ally and hostile forces on a map by combining GIS spatial analysis techniques with common military symbols. MOLE provides a clearer visualization of mission planning and goals for the decisionmaker, and allows a user to import, locate, and display order of battle databases.

Grid Manager

Grid Manager enables the national security intelligence or defense user to create accurate, realistic grids that contain geographic location indicators based on specified shapes, scales, coordinate systems, and units. This program allows the user to create multiple grids, graticules, and borders for such map products as MGRS coordinates and tourist, topographic, parcel, street, nautical, and aeronautical maps.

GOOGLE MAPS

Google Maps is a Web-based service that provides detailed information about geographical regions and sites around the world. In addition to conventional road maps, Google Maps offers aerial and satellite views of many places. In some cities, Google Maps offers street views comprising photographs taken from vehicles.

Google Maps offers several services as part of the larger Web application, as follows:

- A route planner offers directions for drivers, bikers, walkers, and users of public transportation who want to take a trip from one specific location to another.
- The Google Maps application program interface (API) makes it possible for Web site administrators to embed Google Maps into a proprietary site such as a real estate guide or community service page.
- Google Maps for Mobile offers a location service for motorists that utilizes the Global Positioning System (GPS) location of the mobile device (if available) along with data from wireless and cellular networks.
- Google Street View enables users to view and navigate through horizontal and vertical panoramic street level images of various cities around the world.
- Supplemental services offer images of the moon, Mars, and the heavens for hobby astronomers.

References

- Goodchild, Michael F. (July 2009). Giscience and Systems. International Encyclopedia of Human Geography. Pp. 526–538. Doi:10.1016/B978-008044910-4.00029-8. ISBN 9780080449104
- Gis-uses-application: igismap.com, Retrieved 23 March, 2019
- Marwick, Ben; Hiscock, Peter; Sullivan, Marjorie; Hughes, Philip (July 2017). "Landform boundary effects on Holocene forager landscape use in arid South Australia". Journal of Archaeological Science: Reports. 19: 864–874. Doi:10.1016/j.jasrep.2017.07.004
- Levitt, Tom. "High Court battle over London City Airport expansion". The Ecologist. Retrieved 19 May 2011
- Bill Kilday (2018). Never Lost Again: The Google Mapping Revolution That Sparked New Industries and Augmented Our Reality. Harperbusiness. ISBN 978-0062673046
- Google-Maps, definition: techtarget.com, Retrieved 2 June, 2019
- Geographic Information Systems (GIS) Poster, Last modified on 2007-02-22, USGS, Retrieved on 2011-01-16

Permissions

All chapters in this book are published with permission under the Creative Commons Attribution Share Alike License or equivalent. Every chapter published in this book has been scrutinized by our experts. Their significance has been extensively debated. The topics covered herein carry significant information for a comprehensive understanding. They may even be implemented as practical applications or may be referred to as a beginning point for further studies.

We would like to thank the editorial team for lending their expertise to make the book truly unique. They have played a crucial role in the development of this book. Without their invaluable contributions this book wouldn't have been possible. They have made vital efforts to compile up to date information on the varied aspects of this subject to make this book a valuable addition to the collection of many professionals and students.

This book was conceptualized with the vision of imparting up-to-date and integrated information in this field. To ensure the same, a matchless editorial board was set up. Every individual on the board went through rigorous rounds of assessment to prove their worth. After which they invested a large part of their time researching and compiling the most relevant data for our readers.

The editorial board has been involved in producing this book since its inception. They have spent rigorous hours researching and exploring the diverse topics which have resulted in the successful publishing of this book. They have passed on their knowledge of decades through this book. To expedite this challenging task, the publisher supported the team at every step. A small team of assistant editors was also appointed to further simplify the editing procedure and attain best results for the readers.

Apart from the editorial board, the designing team has also invested a significant amount of their time in understanding the subject and creating the most relevant covers. They scrutinized every image to scout for the most suitable representation of the subject and create an appropriate cover for the book.

The publishing team has been an ardent support to the editorial, designing and production team. Their endless efforts to recruit the best for this project, has resulted in the accomplishment of this book. They are a veteran in the field of academics and their pool of knowledge is as vast as their experience in printing. Their expertise and guidance has proved useful at every step. Their uncompromising quality standards have made this book an exceptional effort. Their encouragement from time to time has been an inspiration for everyone.

The publisher and the editorial board hope that this book will prove to be a valuable piece of knowledge for students, practitioners and scholars across the globe.

Index

CPSIA information can be obtained
at www.ICGtesting.com
Printed in the USA
LVHW061638201121
703981LV00003B/53

9 781647 261351